Aspects of the
PHYSIOLOGY OF
CRUSTACEA

UNIVERSITY REVIEWS IN BIOLOGY

General Editor: J. E. TREHERNE
Advisory Editors: Sir VINCENT WIGGLESWORTH F.R.S.
M. J. WELLS   T. WEIS-FOGH

ALREADY PUBLISHED

IN PREPARATION

Aspects of the

PHYSIOLOGY OF CRUSTACEA

A. P. M. LOCKWOOD

Lecturer in Biological Oceanography
University of Southampton

W. H. FREEMAN AND COMPANY
SAN FRANCISCO

First published 1967

Printed and published in Great Britain
by The Aberdeen University Press

Preface

The Crustacea have always been a popular group for physiological research, since they show such diversity in body form and mode of life. To the practical researcher they are particularly attractive because many of the more readily obtainable species are amenable to being kept under laboratory conditions and, as a rule, tolerate simple operations well. Consequently, a considerable amount of physiological knowledge concerning the group is already available, and much of it was most ably summarized a few years ago in the two volumes of *The Physiology of Crustacea* edited by T. H. Waterman (Academic Press). I feel that there is now a need for a more concise book at a level suitable both as a textbook for senior undergraduates and as background reading for postgraduates. In neither of these groups do potential readers usually have much time to devote to a single topic; so I have attempted to provide an outline of the physiology of the Crustacea which can be read at a few sittings and yet give an overall appreciation of the subject.

The coverage is general, but special attention has been given to those aspects where the physiology differs from that of other animal groups and hence the influence of the moult cycle forms a recurring theme. In a book of this size it has been necessary to omit any details not immediately germane to the topic under discussion, but the reference list has been selected so as to enable the reader interested in a particular aspect to find the pertinent literature readily.

Aspects of the physiology of so varied a group as the Crustacea hold such a fascination for the researcher that over the last twelve years I have devoted almost my entire work to that field of study. If this book awakens its readers to this fascination, I shall be well requited.

Many of my colleagues assisted me to avoid errors both of omission and commission. In particular I should like to thank Drs G. M. Hughes, G. Shelton, P. Le B. Williams and D. W. L.

Crompton; they are, of course, in no way to blame for any errors remaining in the text. The Series editor, Dr J. Treherne, has been a never-failing source of advice, and for his patience and guidance I am most grateful. My thanks are due also to Mrs D. Jones, who typed my tapes and manuscript, and to Mrs A. Pitfield, who assisted with the checking of the proofs and in producing the Index. Finally, I must pay tribute to Mr E. A. Smith of Oliver & Boyd for helpful advice and many improvements to the wording of the text, and to Mr J. Christiansen for the high quality of the drawings he prepared for the illustrations.

Southampton, May 1967   A. P. M. Lockwood

Acknowledgements

The illustrations in this book have either been taken direct or adapted and redrawn from a number of books and periodicals. I am indebted to the authors and publishers of these for permission to use the material. Details of the source of each illustration are given in the list on pp. 304 to 309.

Contents

x Contents

1 : Introduction

The many advances in recent years in the study of invertebrates have drawn attention to the fact that all animal groups have much in common in the essential processes of their biochemistry and physiology. Such basic similarities, however, tend to highlight the physiological differences which have occurred in the major groups as a result of specialization along their own lines of structural advance. A major feature of the arthropods is the stout articulated cuticle and its presence has had a profound effect on several aspects of the morphology and physiology of the body. In soft-bodied invertebrates the muscular system is most often in the form of blocks or sheets of muscle fibres. The presence of a mobile but firm exoskeleton has made possible the development of discrete muscles, with all the advantages of dexterity of movement that such a modification confers on the Crustacea. The peripheral nervous system developed to control these muscles is rendered unique in the animal kingdom by the remarkably limited number of motor axons coupled with the presence of nerve axons which can forestall the motor axon impulse and prevent muscular contraction.

Protection of the surface provided by the cuticle has aided in the evolution of forms capable of penetrating onto land or into fresh water from the ancestral home in the seas. Crustacea are, of course, not alone in having evolved this capacity to live in non-marine environments, but it is apparent that the reduction of the permeability of the general body surface due to the presence of the cuticle has been a principal feature in their capacity to maintain a higher blood concentration in dilute media than can soft-bodied forms. The presence of the cuticle has, however, many obvious disadvantages. Indeed some of these are so considerable that it is extremely doubtful, were the arthropods not already in existence, if any zoologist asked to devise the essential characteristics of a novel but viable life form would even consider the possibility of animals with a rigid exoskeleton which has to be replaced periodically in order to allow size increase. Consideration of some of these disadvantages makes it immediately obvious why this should be the case. At the time of moult the animal is, for a period, largely deprived of its exoreceptors, the corneal surfaces of the eye are replaced, the linings of the statocysts and accompanying statoliths are cast off, in some cases with a consequent diminution in the sense of gravity and acceleration, and all the tactile and other sense organs of the general body surface are presumably temporarily incapacitated as the old cuticle separates from the new one beneath it. During this period of sensory impairment the animal's motility is also restricted, because the new cuticle is at first too soft to act as the necessary rigid support for the muscles to pull on. Indeed, as we shall see, special provision seems to have been made in some cases to ensure that muscular movements are not made at inappropriate times; for many decapods possess special inhibitory nerves to the limbs which, since single axons supply antagonistic muscles, seem to serve only the purpose of bringing about a general suppression of such muscle movement as might distort a soft new cuticle. Fore- and hind-gut linings are cast with the remainder of the cuticle, and most of the larger Crustacea cease feeding several days before moult and do not feed again for some days after ecdysis. At this time, however, the basal metabolic rate rises rapidly as reserves are transferred from the hepatopancreas to the cells of the hypodermis secreting the new cuticle. There is even a shift in emphasis from the use of one pathway of glycolysis to another. Prior to moult the calcium level in the blood rises considerably above the level during intermoult as this ion is withdrawn from the old cuticle. At, and shortly after, ecdysis there is rapid intake of water into the blood and cells, resulting in further changes of the blood ionic levels. All these various effects together represent a drastic metabolic fluctuation for which the cells of the body must be adequately prepared. Control over the process of moult is exercised by a complex and well co-ordinated neuro-secretory and humoral system and the effectiveness of the distribution of the hormones is enhanced by a circulatory system which, despite its haemocoelic nature, is so channelled and valved that there is a one-way circulation and a total circulation time of only about one minute even in the larger forms.

A successful solution of the problems associated with cyclical fluctuations in metabolism and physiological function has enabled the Crustacea to take full advantage of the beneficial features of the presence of the cuticle, and the group has proliferated widely. With some 25,000 species it is the third-largest class of invertebrates after the insects and gastropods. Naturally, within such a large group of animals living in a diversity of habitats there is considerable variation in the detailed functioning of the physiological processes as well as in bodily form. The nature of the oxygen carrier of the blood may be cited as a simple example of such diversity. The great majority of Crustacea appear to lack a specific carrier, but where there is one it may be either haemoglobin, as in some members of the sub-class Branchiopoda, or haemocyanin, as in the larger members of the sub-class Malacostraca.

It would be gratifying if adequate data were available on members of all the sub-divisions of the Crustacea to allow us to determine to what extent such differences in physiological functions are group specific and to what extent they are dictated by other factors, such as the nature of the biological niche inhabited by the organisms concerned. Unfortunately the greater part of our knowledge is restricted to the larger members of the Malacostraca, particularly crabs and lobsters, and so, perforce, the great mass of the processes to be discussed in subsequent chapters concerns these animals alone. However, some comparative data are available for other forms and an outline classification of the major groups in the Class Crustacea is therefore included here.

1. Class Crustacea

In common with other arthropods, Crustacea have a ventral nerve cord with a ganglion in each segment, a dorsal brain and haemocoelic blood system with a dorsal heart pumping blood anteriorly. The coelom is reduced and forms only the cavity of

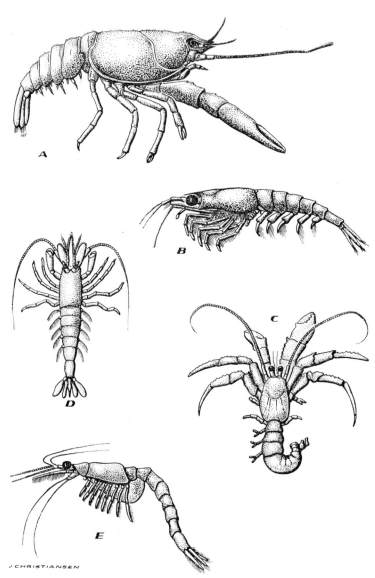

Fig. 1. Various crustacean types: *A. Austropotamobius* (Decapoda, Reptantia). (Redrawn from Borradaile: *Manual of Elementary Zoology*) *B. Thysanoessa* (Euphausidacea). (Redrawn from Hermann Einarsson: *Dana Reports*) *C. Eupagurus* (Decapoda, Anomura). (Redrawn from Calman: *The Life of Crustacea*) *D. Crangon* (Decapoda, Natantia). (Redrawn from Calman, *ibid.*) *E. Praunus* (Mysidacea). (Redrawn from Ritchie: *Outlines of Zoology*.)

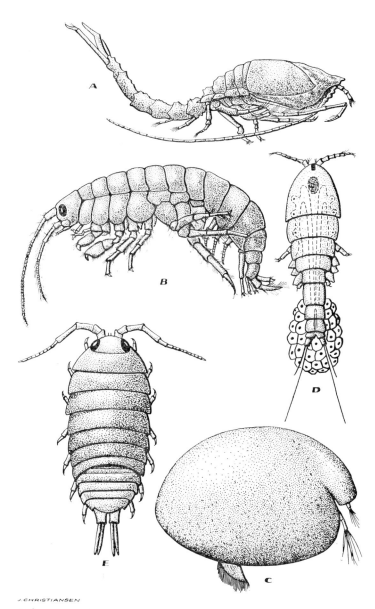

J.CHRISTIANSEN

Fig. 2. Various crustacean types: *A. Diastalis* (Cumacea).
(Redrawn from G. O. Sars: *An Account of the Crustacea of Norway*)
B. Gammarus (Amphipoda). (Redrawn from G. O. Sars: *ibid.*)
C. Cypridina (Ostracoda). *D. Tigriopus* (Copepoda). *E. Ligia*
(Isopoda). (*D* and *E* redrawn from G. O. Sars: *ibid.*)

5

the gonads and the end-sac of the excretory organs. In origin, the excretory organs are presumed to be segmental but in all forms above the Cephalocarida they are restricted to either or both the maxillary and antennary segments. Antennae are carried on the second and third head segments, and mandibles on the fourth segment. Typically, segmented appendages are also carried on all body segments except the first head segment. It is probable that the ancestors of the Crustacea had similar biramous limbs on all the post-cephalic segments and that these limbs carried out the triple functions of locomotion, respiration and food gathering. During the course of evolution, however, the limbs have been modified and specialized for particular functions. The nature of these modifications differs between the various sub-divisions of the group.

Another typical feature of the Crustacea is the development of a carapace. This consists of a fold of skin and its associated cuticle, which overgrows the anterior part (sometimes the whole) of the body. It is not universally present, but when it is developed it usually encloses and protects special respiratory surfaces or facilitates the formation of feeding currents.

Sᴜʙ-Cʟᴀsses

There are eight Sub-Classes:

Cephalocarida	(*Hutchinsoniella*)
Branchiopoda	(Brine shrimps, water fleas, etc.)
Mystacocarida	
Ostracoda	
Copepoda	(*Calanus* and other copepods)
Branchiura	(Fish lice)
Cirripedia	(Barnacles)
Malacostraca	(Crabs, shrimps, woodlice, prawns, lobsters, etc.)

Sub-Class Cephalocarida

This is regarded as the most primitive group of Crustacea, because one of the forms has numerous phyllopodal sub-equal limbs. The type genus *Hutchinsoniella* is the only mandibulate arthropod to retain the primitive eight-segmented limb.

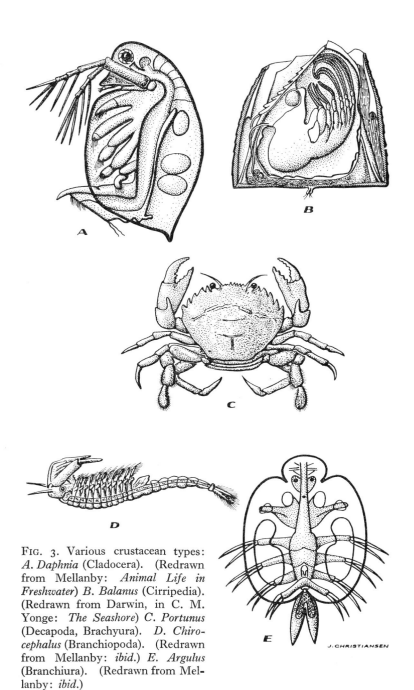

Fig. 3. Various crustacean types: *A. Daphnia* (Cladocera). (Redrawn from Mellanby: *Animal Life in Freshwater*) *B. Balanus* (Cirripedia). (Redrawn from Darwin, in C. M. Yonge: *The Seashore*) *C. Portunus* (Decapoda, Brachyura). *D. Chirocephalus* (Branchiopoda). (Redrawn from Mellanby: *ibid.*) *E. Argulus* (Branchiura). (Redrawn from Mellanby: *ibid.*)

2

7

Sub-Class Branchiopoda

Order Anostraca. No carapace; antennae uniramous; trunk limbs numerous and alike. Eyes compound and stalked (fig. 3*D*).

Order Notostraca. Carapace well developed; eyes compound but sessile; numerous trunk limbs.

Order Diplostraca. Compound eyes closely opposed; antennae large, biramous and used for swimming. Carapace enveloping the trunk region, the limbs of which are modified to produce feeding currents.

Sub-order Cladocera, e.g. *Daphnia* (fig. 3*A*).
Sub-order Concostraca, e.g. *Estheria*.

Sub-Class Mystacocarida

Tiny interstitial forms having some affinities with the Branchiopoda, *Derocheilocaris*.

Sub-Class Ostracoda

Carapace greatly extended to form an enveloping shell, the valves of which can be held closed by an adductor muscle; only two pairs of limbs on the abdomen, e.g. *Cypris*, *Gigantocypris* (fig. 2*C*).

Sub-Class Copepoda

No carapace or compound eyes; no limbs on abdomen. Most are free living but some parasitic, e.g. *Calanus*, *Cyclops*, *Tigriopus* (fig. 2*D*).

Sub-Class Branchiura

Ectoparasitic fish lice. Compound eyes, suctorial mouth, lateral expansion of the head; no limbs on abdomen. Closely related to the Copepoda, e.g. *Argulus* (fig. 3*E*).

Sub-Class Cirripedia

No compound eyes, attached by preoral region and sessile as adult. Usually hermaphrodite, unlike other Crustacea. Carapace enveloping the trunk and usually secreting calcareous plates. Some forms parasitic, e.g. *Balanus*, *Concoderma* (fig. 3*B*).

Sub-Class Malacostraca

Compound eyes; eight thoracic and six abdominal segments, all with appendages. Carapace, where developed, covering thorax only. The largest group of Crustacea, and the one which contains the majority of forms which have been used in physiological studies.

Order 1. Leptostraca. Carapace not fused to thoracic segments, e.g. *Nebalia*.

Order 2. Hoplocarida. Five pairs of subchelate thoracic limbs; carapace fused to three thoracic segments, e.g. *Squilla*.

Order 3. Syncarida. Secondarily lacks carapace, e.g. *Bathynella*.

Order 4. Peracarida. Carapace, if present, fused to four, or fewer, thoracic segments. Brood pouch formed in female by oostegites.

Sub-order. Mysidacea. Carapace well developed, e.g. *Mysis*, *Neomysis* (fig. 1E).

Sub-order. Cumacea. Carapace less well formed, covering only three or four segments, e.g. *Diastylis* (fig. 2A).

Sub-order. Tanaidacea. Carapace small, covering two segments only, e.g. *Apseudes*.

Sub-order. Isopoda. No carapace, body usually dorsoventrally compressed, e.g. *Asellus*, *Ligia* (fig. 2E).

Sub-order. Amphipoda. No carapace, body usually laterally compressed, e.g. *Gammarus* (fig. 2B).

Order 5. Eucarida. Carapace forming a dorsal covering and fused to all thoracic segments.

Sub-order Euphausiacea. Scaphognathite small, statocysts absent (fig. 1B).

Sub-order Decapoda. Well developed scaphognathite and statocysts.

Sub-order Macrura. Abdomen cuticularized, longer than head or thorax; Natantia (fig. 1D), Reptantia (fig. 1A).

Sub-order Anomura. Abdomen not cuticularized, soft (fig. 2C).

Sub-order Brachyura. Abdomen cuticularized, but short and tucked under thorax (fig. 3C).

2 : Osmotic and Ionic Regulation

The cells of all animals must be bathed by a medium of suitable ionic composition and total concentration, if they are to function correctly. The Crustacea, in common with most other major groups of animals, probably evolved in the sea and sea water is sufficiently close to the required composition to maintain the function of tissues isolated from marine Crustacea for short periods. With the evolution of non-marine forms, physiological adjustments have become necessary to ensure the provision of an extracellular fluid suitable for the cells but differing in composition from the medium. Many aspects of the structure and physiology are involved in these adjustments.

1. Marine Forms

Like other invertebrates living in the sea, the majority of marine Crustacea have blood which is isotonic, or almost isotonic, with sea water. In these forms, therefore, there is no tendency for osmosis to cause any gross movement of water across the body surface. One must not, however, be misled by this fact into the belief that marine Crustacea do not need to expend any energy in order to maintain the composition of their body fluids. The blood, though isotonic with the medium and generally similar in ionic composition, differs in detail; and there is an even greater difference between the composition of the cell sap and that of sea water. Work must therefore be done in regulating the normal composition of both extracellular and intracellular fluids.

The most obvious difference between the ionic composition of the blood and that of the medium is that there is generally a somewhat lower level of magnesium in the former[455] (cf. table 1).

Since it has been found that neuromuscular transmission is depressed in *Carcinus, Maia* and *Palinurus*[44, 287, 535] when the concentration of magnesium in the medium bathing the tissues is raised to one and a half or two times the level usually present in the blood, Robertson has suggested that there may be a direct correlation between the maximum rate of movement of individuals

TABLE I

The relative ionic composition of the plasma of some marine Crustacea with reference to a chloride value set at 100 (from Robertson[454a, 455])

Weight Units (gm)

	Na	K	Ca	Mg	Cl	SO$_4$
Sea water	55·5	2·01	2·12	6·69	100	14·0
Dromia	53	2·40	2·04	6·7	100	7·0
Lithodes	58	2·56	2·60	6·7	100	13·0
Hyas	57	2·68	2·52	6·4	100	14·0
Maia	57	2·58	2·90	5·7	100	8·8
Eupagurus prideauxi	61	3·38	3·38	5·3	100	17·7
Eupagurus bernhardus	64	2·88	3·75	3·7	100	19·0
Portunus depurator	61	2·82	2·67	3·3	100	11·1
Cancer	64	2·44	3·00	3·2	100	13·8
Portunus puber	63	3·02	2·92	2·8	100	11·3
Carcinus	62	2·43	2·69	2·4	100	8·0
Squilla	64	2·69	2·59	2·2	100	11·2
Palinurus	63	2·05	2·73	2·0	100	10·4
Pachygrapsus	62	2·24	2·55	1·9	100	7·3
Nephrops	65	1·61	3·02	1·2	100	9·7
Homarus	62	1·75	3·20	1·0	100	4·4

and their blood magnesium-levels. In support of this, he has shown that all the forms capable of rapid movements maintain the blood magnesium at a low concentration. The sulphate ion too is often maintained at lower levels in the blood than in sea water but in this case there is less consistency. The sulphate level in *Eupagurus*, for example, is higher than that in sea water. The ionic composition of some typical forms is given in table 1.

THE MAINTENANCE OF BLOOD IONIC COMPOSITION

Four factors might contribute to the maintenance of differences between the ionic levels in blood and sea water. These are (1) the establishment of a Gibbs-Donnan equilibrium, (2) urinary ion losses, (3) active transport of ions at the body surface, (4) binding of ions by blood proteins. A considerable amount of protein is present in the blood of most species, and hence the possibility arises that the establishment of ion-protein complexes or of a Gibbs-Donnan equilibrium across the body surface may account for the differences in the ionic content of blood and medium. Robertson[455] has investigated the possibility of the differences having arisen in this way by dialysing the blood of several forms against sea water through a collodion membrane. The final dialysate differed from the normal blood in the case of all nine species studied. The general interpretation of this observation is that active transport mechanisms are involved in the maintenance of the normal body fluid composition. Not all the ions whose levels differ before and after dialysis are necessarily actively transported. The distribution of charged particles across membranes is governed by both concentration and electrical gradients. The electrical gradients are determined by a complex of factors including the concentration gradients of ions across the membrane and the relative permeability of the membrane to different types of ion. Without knowledge of the electro-chemical gradients across different membranes or body walls it is not practicable to compute which ions are in electro-chemical equilibrium and which are actively transported. A further complication is introduced by the fact that ions are continuously being lost from the blood in the urine. If the rate at which the different ions are replaced varies, then this continuous concentration will itself affect the ionic composition of the blood.[397] It is possible that in the more permeable forms the colloid osmotic pressure of the blood proteins serves to bring into the body sufficient water and ions to replace the loss in the urine; but it seems likely that in most cases this is supplemented by active uptake of ions from the gut or across the body surface. Active transport is considered in more detail on pp. 30 ff.

THE CELL SAP

Though few determinations have been made of the osmotic pressure of the cell sap in Crustacea, it seems that it is very close to being isosmotic with the blood.   For example, careful measurements on the abdominal flexor muscles of *Nephrops*,[457] indicate that the osmotic pressure differs by only 2% from that of the plasma.   Even this small difference may be in part accounted for by tissue breakdown prior to the determination, as was suggested by Shaw[478] to account for a difference of 7 – 8% between the measured blood osmotic pressure and that of the sap of the carpopodite extensor and flexor muscles in *Carcinus*.

A relatively small proportion of the total osmotic pressure of the cells is accounted for by inorganic ions.   In the carpopodite of *Carcinus*, these ions, plus organic phosphates, are responsible for only a little over one-third of the osmotic pressure.

TABLE 2

mM/kg *muscle-fibre water*

	K	Na	Cl	Mg	Ca	ATP	AP	mo sm/kg
Nephrops abdominal flexor	188	24·5	53·1	20·3	3·72	13·2	117·8	1037
Carcinus carpopodite	146	54	53	17	5	9	82	983

Nephrops data from Robertson,[457] *Carcinus* from Shaw.[478]

The muscles of *Nephrops* and *Carcinus* show considerable similarity in their ionic composition.   Both these tissues, however, are muscles capable of fast contraction and it is likely that other tissues will have different ionic compositions.   For instance a determination of the potassium content of crab nerve indicates a concentration of 330–440 mE/kg, a level some two to three times that of muscle.[291]   Such differences between the composition of different tissues are well known in the vertebrates, and Potts[419] has also found differences in the ionic composition of the slow and fast contracting muscle of *Mytilus*.

The remainder of the osmotic pressure of the muscle cell sap is produced by small organic molecules of which the most important (quantitatively) are amino-acids, betaine, and trimethylamine oxide.

Soluble carbohydrates are also present in the tissues, but they account for a much smaller part of the osmotic pressure.   In the lobster, *Homarus*, glucose and other carbohydrates make up some

3·4 mM/kg tissue water and lactic acid about 9·1 mM/kg tissue water.[48]

The calculated osmotic activity contributed by the various components of *Nephrops* muscle is found to exceed the measured osmotic pressure of the sap. However, as the cell sap extruded when the cells are squeezed has a different composition from that given by analyses of the whole muscle fibres, Robertson[457] concludes that a proportion of the ions in the tissue must normally be 'bound' to proteins. Such bound ions would not contribute to the total osmotic pressure of the sap. On the basis of analyses of the muscle press juice and the whole fibre, he calculates that chloride

TABLE 3

mM/kg *tissue water*

	α-amino N	Betaine	T.M.O.	
Nephrops	476	65·7	59	Robertson[457]
Carcinus	434*	93	90	Shaw[478]

*Not including arginine.

occurs in the fully ionized state, but that all the calcium, 82% of the sodium, 60% of the magnesium and 25% of the potassium are osmotically inactive in the cell. No comparable determinations for ion binding are available for other Crustacea, but there is some evidence of a similar phenomenon in vertebrate tissues.[109, 496]

The ionic concentrations inside the cells are very different from those of the extracellular fluid. Of the more mobile ions potassium is more concentrated inside the cell than outside, whilst the reverse is true of sodium and chloride. Consequently concentration gradients are present which would be expected to dissipate the concentration difference by diffusion if the ion levels were not actively regulated. In fact interference with the metabolic activity of cells does result in a loss of potassium and a gain of sodium, chloride and water.

As is the case in the tissues of other animals the low sodium level of *Carcinus* cells is maintained by the continuous active extrusion of sodium.[475, 477] However, a positively charged ion such as sodium cannot be extruded from a cell unless either a negatively charged ion accompanies it or another positive ion moves into the cell. The evidence at present available is compatible with the assumption that extrusion of sodium is accompanied by an uptake of potassium into the cell. As potassium is at a higher concentration inside the cell than outside, it tends to diffuse out. However, a large part of the negative charge inside the cell is contributed by organic substances, which cannot readily escape. Hence the tendency for potassium to diffuse out causes a separation of charge, which is manifested as a potential across the cell membrane. When of sufficient magnitude, this membrane potential counteracts the diffusion gradient and prevents any further net loss of potassium. The balance point, reached when the driving forces produced on the potassium ion by the electrical and chemical gradients are equal and opposite, is defined by the Nernst equation :

$$E = \frac{RT}{zF} \ln \frac{K_1}{K_0}$$

where E is the membrane potential, R is the gas constant (8·2 joules per mol./degree absolute), T is the absolute temperature, z the valency of the ion, F is the Faraday (96,500 coulombs/mol.) and K_1 and K_0 are the potassium activity inside and outside the cell. At a first approximation the potassium activities are represented by the internal and external concentrations.

The establishment of such a 'diffusion potential' across the cell membrane naturally influences the movements and final concentrations of other ions, since these are governed by the total electro-chemical gradient and not just by their diffusion gradient. If there were only one other ion in the system, e.g. chloride, and if this ion were in purely passive equilibrium across the cell membrane, it would be expected that the following relationship would hold:

$$\frac{RT}{zF} \ln \frac{K_1}{K_0} = E = \frac{RT}{zF} \ln \frac{Cl_0}{Cl_1} \text{ or } \frac{K_1}{K_0} = \frac{Cl_0}{Cl_1}$$

Such a relationship would also hold good in a multi-ionic system if the cell membrane had a very low permeability to ions other than potassium and chloride. In the case of some vertebrate tissues, such as the frog sartorius muscle, this equation is applicable.[254]   In *Carcinus* muscle the potassium ratio is 12 and that for chloride 10·3.[475, 477]   Allowing for the binding of some of the internal potassium, as in the muscles of *Nephrops*, these two ratios would also be very similar and the equation applicable.   In *Nephrops* itself, however, the ratios are quite unequal, that for potassium being 21·9 and that for chloride 9·9.[457]   Here it is possible that the chloride is in electro-chemical equilibrium but the potassium is not.   The probable explanation of this observation is that the membrane potential is not determined solely by the potassium ion in membranes where there is a finite permeability to other ions.   A better description of the membrane potential is that given by the modified Goldman equation[119]:

$$E = \left(\frac{RT}{F} \ln \frac{pK[K_o] + pNa[Na_o]}{pK[K_i] + pNa[Na_i]}\right) = \frac{RT}{F} \ln \frac{[Cl_i]}{[Cl_o]}$$

if the chloride is in passive flux equilibrium.   When potassium is not in electro-chemical equilibrium across the cell membrane, it may be concluded that its concentration in the cell, like that of sodium, is regulated by an active process.

A difficulty in relating the membrane potential of crustacean muscles to this equation was raised by the observation[477] that the tracer flux of sodium into isolated fibres was greater than that for potassium.   This would suggest that the permeability of the membrane is greater to sodium than to potassium and hence that the membrane potential could not be related to the diffusion potentials.   It is probable, however, that there are deep intuckings of the surface in which sodium, strictly outside the cell, may lie and yet participate in what appears to be an exchange of cell sodium, since it has been observed that when a micro-electrode is pushed at an angle into a muscle fibre the membrane potential comes, goes and reappears as the tip is pushed deeper.[178]

The maintenance of different ionic concentrations across the cell membrane plays an important role in regulating the volume of cells.[331, 332]   Some of the total osmotic constituents of the cell are unable to diffuse freely across the cell membrane, whilst the major part of the blood osmotic pressure is accounted for by sodium and chloride, which can penetrate into the cell.   In the absence of corrective measures the colloid osmotic pressure of the cell contents would therefore result in the entry of sodium, chloride and water.   Cellular swelling from this cause is believed to be prevented by the establishment of a ' Double Donnan equilibrium '.[331, 332]   It is argued that the continuous extrusion of sodium

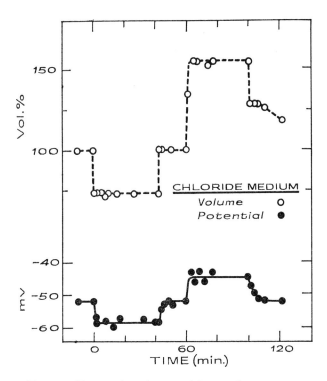

Fig. 4.   Changes in volume and in membrane potential of a crayfish muscle fibre exposed first to a medium made hyperosmotic by addition of 165 mM NaCl, returned to physiological saline, and then exposed to a hyposmotic medium.

Note the hyperpolarization associated with shrinkage and the depolarization related to swelling.   (From Reuben, Girardier and Grundfest.[440])

from the cell as fast as it diffuses in down the electro-chemical gradient creates a situation where the cell membrane is, in effect, rendered impermeable to this ion.   Net chloride entry into the cell is prevented by the membrane potential. In consequence the blood ion concentration can off-set the osmotic effect of the non-penetrating organic molecules within the cell. Naturally, osmotic equilibrium breaks down if sodium is not continuously extruded. Osmotic balance is disturbed if the concentration of the blood is changed (see pp. 23ff.); but it is also upset if the potential across the cell membrane is varied. Thus it is found that the volume of *Procambarus* muscle cells alters when the membrane potential is artificially depolarized or hyperpolarized (fig. 4).

Such changes are presumed to indicate that water movements can occur by electroosmosis across the cell membrane, as it does in the case of artificial charged and selectively permeable membranes of comparable properties. Study of the artificial membranes suggests that the rate of water movement by electroosmosis would be expected to depend on the sign and number of fixed charges on the membrane, the degree of ionic selectivity of the membrane, the concentration of the ionic species present on either side, and the potential difference across the membrane. Cell volume cannot be expected to remain constant, therefore, unless the various blood and cellular parameters are accurately regulated. This point has particular relevance when one considers the effects on the blood concentration of transferring marine forms to dilute media.

2. Marine Crustacea in Dilute Media

When marine Crustacea are placed in diluted sea water there is a general loss of ions across the body surface and in the urine, the cellular ion content is disturbed and the cells take up water by osmosis from the blood. If these effects are at all extensive the animals die. Consequently sub-littoral marine Crustacea will not tolerate any marked dilution of their medium. The crabs *Maia* and *Emerita analoga* will survive in 80% and 75% sea water respectively, but these levels represent the limits to which such purely marine forms can be adapted.[160, 227] However, from the representation of the numerous crustacean orders, families and even genera in both the sea and fresh water, it is apparent that the evolution of the mechanisms necessary to enable the colonization of dilute media has occurred many times.

3. The Blood Concentration of Euryhaline Forms in Relation to the Concentration of the Medium

All Crustacea capable of tolerating a range of salinities (euryhaline forms) tend to maintain their blood hypertonic to the

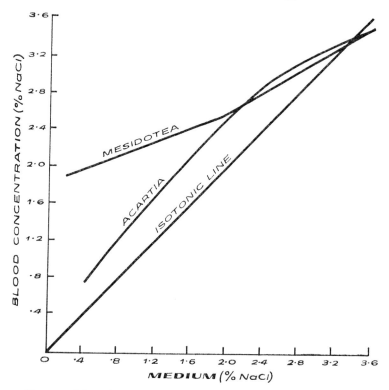

FIG. 5.   The relationship between blood concentration and concentration of the medium for *Mesidotea entomon* (Cl) and *Acartia tonsa* (osmotic pressure).   (*Mesidotea* values from Lockwood and Croghan,[343] *Acartia* values from Lance.[320])

medium, at least when the latter is dilute.   The degrees of hypertonicity maintained vary widely.   For example, in fig. 5 it can be seen that the isopod *Mesidotea entomon* regulates the blood so that there is a much greater gradient of concentration between blood and medium at low salinities than in the euryhaline copepod

Acartia tonsa.[319, 320]   Most of the other brackish water Crustacea so far investigated, which are hypertonic regulators, have blood levels which lie approximately between these extremes, though the extent of their salinity tolerance varies.   Thus the amphipod

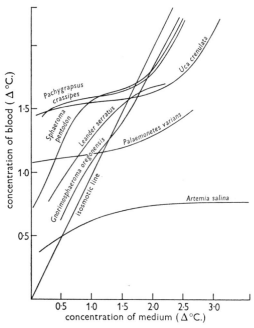

FIG. 6. Various levels of hyposmotic and hyperosmotic regulation, showing the range between almost homoiosmotic regulation and almost complete conformity with the medium. (Data from the following authorities: *Artemia salina* from Croghan, *Palaemonetes varians* and *Leander serratus* from Panikkar, *Sphaeroma pentodon* and *Gnorimosphaeroma oregonensis* from Riegel, *Uca crenulata* and *Pachygrapsus crassipes* from Jones.)

Gammarus duebeni, which has a blood concentration intermediate between that of *Acartia* and *Mesidotea*, tolerates more dilute media than the former.   The forms considered above are hypertonic to the medium only when the latter is below a given concentration. Above this level, the blood and media approximate to isotonicity. A number of species can, however, maintain the blood not only hypertonic to dilute media but also hypotonic to more concentrated media. These include species which are semi-terrestrial or from waters of fluctuating salinity. *Pachygrapsus crassipes*,[227] *Artemia salina*,[115] the isopods *Gnorimosphaeroma oregonensis* and *Sphaeroma pentodon*[443] the mysid *Neomysis integer*[428] and the prawns *Palaemonetes varians*[391] and *Palaemon serratus* (fig. 6).[403] Related animals from more stable environments may not have the capacity to maintain the blood concentration at a level less than that of the medium. Thus the marine prawns *Pandalus montagui* and *Pandalina brevirostris* are isotonic with sea water,[393] whilst the American prawn *Palaemonetes paludosus* from fresh water and dilute brackish water is unable to regulate its blood hypotonic to saline solutions more concentrated than the usual blood level.[142]

A number of different processes are involved in the regulation of the blood concentration at a level differing from that of the medium. These include (1) reduction in the permeability of the body surface, (2) toleration at the cellular level of variations in blood concentration, and (3) the ability to transport inorganic ions against the concentration gradient across the body surface. Associated with this last has been the evolution of an ion uptake mechanism with a high affinity for ions. A few forms also possess the ability to conserve ions within the body by the production of urine hypotonic to the blood.

REDUCTION OF THE PERMEABILITY OF THE BODY SURFACE

Most Crustacea living in water of low salinity maintain their blood hyperosmotic to the medium by actively transporting salts into the body. The rate at which they have to do this, and consequently the amount of work that has to be done in regulating the blood concentration, depends on the rate at which ions are lost from the body, in the urine and across the surface. The rate of loss depends on the permeability of the body surface to ions and water and on the gradient maintained between the blood and medium. In marine forms the permeability of the surface is usually high, partly perhaps because no premium is placed on impermeability when the blood concentration is so similar to that of the medium, but possibly also because it may be an advantage to have a high permeability in order to obtain water

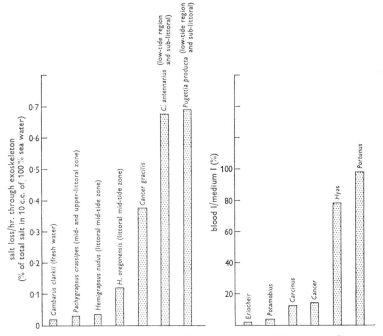

FIG. 7.   The permeability of the exoskeleton of Crustacea from different environments, showing that there is a tendency for permeability to be restricted in animals from dilute media or semi-terrestrial habitats.

Left: The relative permeability of the exoskeleton; the values are the means of three determinations with a gradient of 50% sea water across a portion of cuticle.   (Permeability data from Gross[228]; data on shore distribution from Ricketts and Calvin (1948).)

Right: Ratio of iodine in the blood to that in the medium 2½ hours after placing various species in sea water containing sodium iodide.   (Data from Nagel.[381])

for urine production.   In the lobster, *Homarus*, only about a fifth of the volume of the urine can be attributed to water taken in from the gut[72]; hence, a considerable proportion must be taken in across the cuticle.   As anuria and oliguria in this animal are associated with low blood protein levels and as anuric lobsters often resume urine production after the injection of serum it seems reasonable to suppose that the colloid osmotic pressure of the blood proteins plays an important role in water uptake from the medium. Most of the water intake by intermoult *Carcinus* is also across the surface.[456]

The permeability of the body surface of brackish-water, fresh-water and littoral species is lower than that of marine species. The degree of permeability tends to be related to the habitat in which the animal lives; sub-littoral forms are more permeable than brackish-water species, which in turn are more permeable than fresh-water forms (fig. 7).[227, 381] It is doubtful if the increase in permeability of the surface can be attributed solely to changes in the permeability of the cuticle, as the cuticle over the gills of *Eriocheir* is extremely thin, unlike that in the much more permeable *Maia*.[539] It seems likely, therefore, that the epithelial cells of the gills themselves are in part responsible for the control of permeability. In the case of *Carcinus* the animal does not appear to be able to effect changes in the permeability of the cuticle when the concentration of the medium is changed but it has been found that the gills of the fresh-water crayfish *Astacus leptodactylus* become more permeable to water as the concentration difference between blood and medium declines.[35]

The rate of passive movements of ions across membranes is governed by the electrical gradients as well as the concentration gradients. For example the presence of a 40 mV potential across the body wall of an animal (inside negative) would be expected to double the passive influx and halve the passive efflux of monovalent cations.[419b]

CELLULAR ADAPTATION TO REDUCTION IN
THE CONCENTRATION OF THE BLOOD

The maintenance of the blood concentration at a level higher than that of the medium is made feasible by the active uptake of ions together with the decrease in permeability described above. However, the body surface cannot be made completely impermeable to salts and water without giving rise to an unacceptable restriction of gaseous exchange. Hence, if the blood concentration were to be kept at a level very much higher than that of the medium, the amount of work which would have to be done in transporting ions to replace those lost from the body might be prohibitive. For example, it has been calculated that, were *Carcinus* to be able to live in fresh water and maintain in that medium the concentration of blood that it has in 40% sea water, then some 30% of its total available metabolic energy would be used in transporting ions across the body surface, even if only the minimum thermodynamic energy expenditure was made.[480]

The decrease in blood concentration which occurs in all brackish-water animals as the concentration of the medium falls, serves to limit the energy expenditure on osmoregulation. However, none of the Crustacea so far studied are able to keep their cells hypertonic to the blood. Therefore, if the blood concentration falls, provision must be made to adjust the internal osmotic concentration of the cells, or they will take up water from the blood. Any gross water movements between extra-cellular and intra-cellular compartments must be avoided if cellular disruption or restriction of circulating volume is not to occur. To give emphasis to the point that mere toleration of cellular swelling is not practicable, Croghan (unpublished) has considered the water shifts which would be expected in a hypothetical marine form incapable of cellular osmotic adjustment if it were suddenly transferred to a dilute medium. He shows that, if the cells are regarded as perfect osmometers (and crayfish muscles in hypotonic media approximate to this condition[440]), then the product of the cell volume and the blood concentration is a constant. Hence, on dilution of the blood, water uptake by the cells would be expected to reduce the extra-cellular volume to zero when

$$\frac{C_b}{C_{b_0}} = \frac{V_{a_0} - V_{b_0}}{V_{a_0}}$$

where C_b is the blood concentration, C_{b_0} is the initial blood concentration, V_{a_0} is the initial volume of the whole animal, V_{b_0} is the initial volume of the extra-cellular space. For example, a form with an initial blood volume of 50% of its total volume would have a zero blood volume when its blood concentration had fallen to 50% of the initial value ; one with an initial cell volume of 70% of the total would have no blood when the concentration had fallen by 30%.   Obviously no real animal could tolerate a decrease in blood volume sufficient to interfere with the circulation, let alone entirely eliminate the blood, so it is clear that cellular osmotic adjustment must occur if a decrease in blood concentration is to be tolerated.   The measure of adjustment varies from

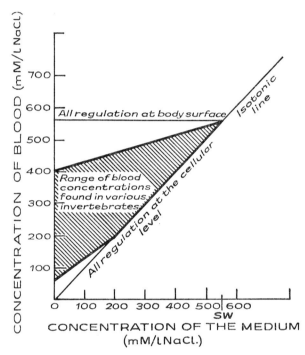

FIG. 8.   The relation between osmotic regulation at the body surface and at the level of the general cells of the body in brackish-water animals maintaining various gradients of concentration between their blood and the medium.   (From Lockwood: *Animal Body Fluids and their Regulation.*)

one species to another.   In a form whose blood concentration remains isotonic with the medium as the latter changes in concentration (fig. 8), all the adjustment must be at the cellular level. If the blood concentration were to be kept constant as the medium was diluted, then all the adjustment would be to mechanisms responsible for maintaining the blood ionic level, and no change would be necessary at the cellular level.   In fact, as we have seen, no crustacean is able to keep the blood absolutely constant during wide changes in the concentration of the medium, though a few, such as *Artemia*,[115] some prawns[391] and mysids,[428] can maintain a fair degree of constancy (fig. 6).   Consequently the ability is needed both to vary the rate of ion uptake to match differences in rates of loss as the concentration gradient varies, and also to vary the cell osmotic activity as the blood concentration changes.   The extent to which reliance is placed on cellular adjustment or on ion transport depends on the size of the osmotic gradient maintained between blood and medium.

CELLULAR OSMOTIC ADJUSTMENT

Since inorganic ions contribute only about one third of the osmotic activity of the muscle fibres of marine forms such as *Nephrops* and *Carcinus* they cannot be solely responsible for wide changes in cell osmotic pressure.   The degree of ionic contribution to such changes is further restricted by the fact that during moderate dilution of the blood the ratio of the concentration of blood potassium to cell potassium is kept substantially unaltered.[477] Thus the membrane potential of *Carcinus* muscle is kept practically constant while the osmotic concentration of the blood is varied in the range 580 to 200 mM/l NaCl.[476]

Much of the osmotic adjustment of cells is due to variations in the concentration of amino-acids, though other organic substances, such as trimethylamine oxide and taurine, also contribute. When a euryhaline form such as *Carcinus*[150, 478] or *Eriocheir*[149] is transferred from sea water to a dilute medium, there is a marked drop in the free amino-acid concentration of the cells (table 4).[478] This process has been extensively studied by Florkin and his associates,[50, 82, 149, 150, 151, 152, 197] who have shown that it is of general occurrence in brackish-water forms.   *Arenicola*, *Nereis diversicolor*, *Perinereis cultrifera* and *Asterias* in addition to the crabs and prawns such as *Palaemon serratus* and *P. squilla* all reduce their muscle amino-acid concentration to some extent when put into dilute media.   The process is fully reversible. Thus, when *Eriocheir sinensis* is transferred back to sea water after acclimatization to fresh water it regains a high concentration of amino-acids in its cells.[49, 153]

TABLE 4 (from Shaw[478])

	Mean concentn. in muscles of crabs from 100% sea water mM/kg fibre water	Mean concentn. in muscles of crabs from 40% sea water mM/kg fibre water	Concentn. for 40% crabs calculated from change in water content mM/kg fibre water	Loss not accounted for by change in water content mM/kg fibre water
α-amino-N compounds excluding arginine	436	191	355	164
Trimethylamine oxide	90	58	73	15
Unidentified N compounds	93	66	76	10
Water content (%)	74	77·8	—	—

TABLE 5 (from Bricteux-Gregoire et al. [49])

Differences in the concentration of osmotically active constituents of Eriocheir *muscles on transference from fresh water to sea water.*   mM (*or* mgm *ions*)/kg *water*

	Animal from fresh water	Animal from sea water	Difference
Inorganic ions			
Cl, Na, K, Mg, Ca	203·5	483·4	279·9
Total amino-acids	157·9	340·7	182·8
Taurine	17·3	20·7	3·4
Trimethylamine oxide	47·6	74·8	27·2
Betaine	17·8	13·9	−3·9

Most of the common amino-acids have been found to occur free in the muscles of Crustacea, but the major contribution to changes in total concentration is made by glycine, proline, glutamic acid (+ glutamate) and alanine, all of which are present in considerable quantities in the amino-acid pool.   Arginine, which is also well represented, plays no role in osmotic adjustment, possibly because much of it is present primarily in the form of the phosphagen arginine phosphate as an immediate energy store (table 6).

TABLE 6 (from Duchateau-Bosson and Florkin[153])

Free amino-acids in the muscles of Eriocheir sinensis *in fresh water or adapted to sea water for nine weeks:* mg/100 gm *fresh muscle*

	Animals from fresh water	Animals from sea water	Difference
Alanine	412·1	698·1	286
Arginine	614·2	470·5	−143·7
Aspartic acid	43·3	115·5	72·2
Glutamic acid	252·3	462·2	209·9
Glycine	718·1	1047·3	329·2
Histidine	5·3	16·0	10·7
Isoleucine	9·9	35·5	25·7
Leucine	10·2	45·5	35·3
Lysine	40·5	63·8	23·3
Methionine	9·5	37·6	28·1
Phenylalanine	4·0	10·9	5·9
Proline	497·2	1600·1	1102·9
Threonine	14·7	64·7	50·0
Tyrosine	3·2	9·1	5·9
Valine	15·4	47·3	21·9

The relative importance of the amino-acids involved in osmotic adjustment is somewhat variable, even between species of Crustacea. In other groups even more striking differences appear. Thus, the limited adjustment of which *Asterias* is capable is almost solely the result of a change in the concentration of glycine and taurine.[278] In *Arenicola* glycine, alanine and, to a lesser extent, glutamic acid play the major part in the adaptation of the cells to osmotic change. Proline is not varied.[151] As a result of changes in the amounts of osmotically active substances, cell volume is quite well controlled in good regulators. Thus, *Carcinus* shows only about a 3% change in muscle water when transferred from 100% to 40% sea water,[478] whilst there is a change of about 5% in *Pachygrapsus* put from 150% to 50% sea water.[235] Not very much is yet known of the way in which changes in the amino-acid concentration are brought about or the nature of the stimulus which evokes the change. There is only a small amount of peptide in the cells of Crustacea,[478] so it does not seem probable that the breakdown and reformation of special peptides can be the basis of the mechanism. Three remaining possibilities are (1) that proteins are degraded or reformed from the amino-acid pool, (2) that the amino-acids are formed *de novo* as required, (3) that the amino-acids are taken up from the blood when required.

The major contributors to the amino-acid regulation, glycine, proline, glutamic acid and alanine, are all in the category of non-essential amino-acids. Hence it is possible that they can be formed by amination of products of glycolysis or the tricarboxylic acid cycle, and deaminated when no longer required.[197] Some support for this last being the method actually used is provided by the observation that the nitrogenous waste secreted by *Eriocheir* is markedly increased for a short period after transference of the animal from sea water to fresh water, and conversely is temporarily decreased when the reverse transference is made.[277] The same phenomenon is found when *Carcinus* is transferred from one concentration to another.[382] Nevertheless, the possibility that part of the changes in the amino-acid concentrations results from variations in protein metabolism cannot yet be ruled out. Potassium deficiency accompanied by sodium penetration into muscles is associated with interference with protein metabolism in rats,[379] and it is therefore possible, though unproven, that a similar rise of sodium in the tissues of Crustacea might result in an accumulation of amino-acids in the cells. Preliminary experiments do indeed seem to indicate that the concentration of inorganic ions within the cell, rather than the total osmotic pressure, may be responsible for determining the amino-acid level. Nerves taken from an *Eriocheir* adapted to fresh water and put into a dilution of sea water isotonic with the blood showed a much lower level of amino-acids after twenty-four hours than did a similar batch of nerves put into 100% sea water. However, nerves put into 50% sea water, plus enough sugar to make the medium isotonic with sea water, showed a decrease in the intracellular amino-acid concentration.[466] As a result of these experiments it may be concluded (1) that it is not the osmotic pressure of the medium *per se* which governs the intracellular amino-acid levels, (2) that the amino-acids have their origin inside the cells and are not taken up from the medium, and (3) that changes in the amino-acid levels can be effected by the cells in the absence of hormones.[466]

One of the cell enzymes most likely to be involved in changes in amino acid concentration is glutamate dehydrogenase. The reaction it catalyses is the interaction of glutamic acid with NAD to give $NADH_2$ plus α-oxo-glutaric acid (α-keto-glutaric acid). The equilibrium position of the reaction favours the formation of glutamic acid. Thus activation of this enzyme by inorganic ions could bring about an increase in the level of free glutamic acid. If transamination with the appropriate keto-acid substrates then occurred a general rise in free amino acids might be brought about.

Inorganic ions have indeed been found to influence the enzyme activity.[266b, 466b] Examination of the glutamate dehydrogenase from *Homarus* and *Carcinus* shows that it can be activated by sodium chloride and other inorganic ions; an interesting observation since the NaCl content of cells would be expected to rise as the concentration of the medium bathing them is increased.

Further evidence that this enzyme participates in amino acid synthesis is provided by incubation experiments. When crab hepatopancreas in incubated with ^{14}C-acetate, the label rapidly appears in glutamic acid.[266c] The proportion of the label present in glutamic acid is smaller when the external medium is 50% sea water than when it is 100% sea water.[266b]

Fresh-water animals, such as the crab *Potamon niloticus* and crayfish *Astacus astacus*, show some variation in cell amino-acid level when they are transferred to more saline media but the concentration differences are less than in brackish-water species, and these animals are not able to survive in full-strength sea water.[152, 481]

THE ACTIVE UPTAKE OF INORGANIC IONS

When in their normal environment, most species of Crustacea can replace the ions which they lose across the body surface and in the urine by actively taking them up from the water. There are however a few species, such as *Branchipus*[309] and *Chirocephalus*,[392] which are unable to balance loss and uptake of ions unless they are fed.

The various inorganic ions are lost from the body at different rates; for instance, the crayfish *Austropotamobius pallipes* (*Astacus fluviatilis*) loses chloride at only about 70% the rate at which sodium is lost.[483]   Consequently, independent mechanisms are necessary for the uptake of the different ions if the normal ionic ratios of the blood are to be maintained. *Eriocheir* can take up chloride from NaCl, KCl, NH_4 Cl and $CaCl_2$[414] and sodium from NaCl, Na_2SO_4[356] and $NaHCO_3$.[414]   Sodium and chloride ions are usually taken up together, but if the ionic balance of the body is disturbed then one or other is taken in more rapidly.   Thus, if *Eriocheir* is depleted of chloride by being kept for a time in $NaHCO_3$, it will subsequently, when it is returned to NaCl, take up chloride faster than sodium.   When the crayfish, *Austropotamobius*, is replacing a loss of both sodium and chloride, it will take up the former faster than the latter, so that the uptake of sodium may be completed while that of chloride is still continuing.[483]   Although the mechanisms for sodium and chloride uptake thus seem to be quite distinct, it is thought that the chloride transporting system may be dependent on that transporting sodium, in so far as the absolute concentration of the blood is set by the sodium mechanism and the chloride mechanism acts to keep a constant Na/Cl ratio in the blood.[483]

When sodium is being taken up without chloride, or chloride without sodium, a similar quantity of ions with the same charge must be liberated from the body if electrical neutrality is to be maintained.   It seems probable that when sodium is being taken up alone ammonium or possibly hydrogen ions are liberated as counter-ions[484, 485] and that bicarbonate is released in exchange for chloride.[414]

Little is known of the details of the mechanism by which sodium is transported but a sodium-potassium sensitive A.T.P.ase, found in the gills of the land crab *Cardiosoma guanhami*, may play some role in the process.[427b]

The maintenance of the blood concentration and ionic ratios requires not only the basic means of transporting ions but also some method of regulating the rates of transport.   The factors known to influence the rate of ion uptake are (1) the concentration of the blood, (2) the concentration of the medium, and (3) the temperature.   In vertebrates the blood volume also influences the mechanisms responsible for retaining sodium in the body; preliminary experiments suggest that the same may also be true of Crustacea.

The Influence of the Blood Concentration

When the concentration of the blood of the crayfish is caused to fall below its ' normal ' level the rate of active uptake of sodium is increased.   A fall in concentration of only some 1–2% results in some increase above the level required to balance the usual sodium loss from the body, but the maximum rate of transport

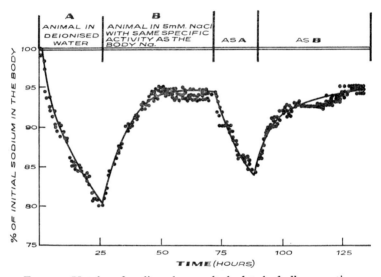

FIG. 9.   Uptake of sodium by a salt-depleted *Asellus aquaticus*, showing the decline in rate of uptake as the sodium level in the body rises towards the normal level.

is not reached unless the blood concentration is some 6–8% below ' normal '.[482]   A similar activation of the mechanism responsible for the uptake of sodium has been found in other Crustacea after a lowering of the blood concentration.[66, 338, 479] Such an increase in the rate of uptake tends to restore the blood concentration to the ' normal ' value and, as it approaches this level, the rate of uptake progressively declines back to the usual maintenance rate (fig. 9).   If the blood concentration is lowered too far, the increased uptake may not be as well marked as normal — presumably because of tissue damage.[66]

As a matter of convenience for further discussion the process of activation of the transporting system will be regarded as involving an increase in the number of transporting sites, though there is no direct evidence that this is actually the case.

The Effect of the Concentration of the Medium

When the medium is diluted there is no change in the rate of sodium uptake by the crayfish *Austropotamobius pallipes*, until the

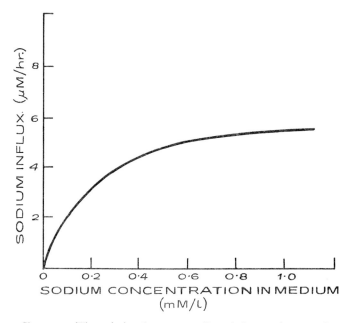

FIG. 10. The relation between sodium influx and external sodium concentration in a single crayfish, showing that at concentrations of the order of 1 mM/l the influx tends towards an asymptotic value. (From Shaw.[482])

external concentration of NaCl is less than about 1 mM/l. At concentrations below 1 mM/l NaCl, the rate of uptake declines (fig. 10) in a manner similar to that described by the

Michaelis-Menten equation for enzyme-substrate interactions. Such a relationship might be expected if the sodium were being transported by a carrier system which tended towards full saturation at about 1 mE/l Na. As the transporting system is then nearly saturated, the rate of uptake is not increased if the concentration of the medium is raised above this level. In fact, very high concentrations may even inhibit the transport system; so the uptake rate falls again. The value for half saturation, which can be calculated with greater accuracy than that for full saturation, is at about 0·2 – 0·25 mE/l in the crayfish,[482] but varies in other species as we shall see (p. 35).

The total uptake of sodium is, obviously, the product of the rate of uptake per site and the total number of sites that are active. Hence, if the concentration of the medium is reduced below the level required to saturate the sites, the total uptake will fail to balance the normal loss of sodium from the animal: and the blood concentration will therefore fall. However, the fact that the sites may not be fully saturated at low concentrations of the medium does not mean that the animal will continue to lose salts until it dies. It will show a net loss of salt only until such time as the drop in blood concentration is sufficient to result in the activation of enough additional sites to balance the loss. The blood concentration can then be maintained constant, but only at a sub-normal level.

It may be presumed that the maintenance of the blood at a sub-normal concentration is disadvantageous; and in fact those Crustacea which inhabit dilute media have evolved active uptake systems with a much higher affinity for sodium than forms which live in more saline environments (Table 7).

The evolution of a high affinity uptake system can be seen in the Glacial relict isopod, *Mesidotea entomon*. This animal lives throughout the Baltic in saline waters and also in a number of Swedish lakes. The Baltic race has a value for half saturation for sodium transport of about 9·0 mM/l Na whilst that of the fresh-water race living in Lake Mälaren is only 2·75 mM/l. Mälaren, originally part of the Baltic, became a fresh-water lake about the twelfth century. It may be expected that the *Mesidotea* in Lake Vättern, isolated approximately 7000 years ago, will have advanced still further in the same direction.[343b]

TABLE 7

The concentration of the medium required to half saturate the transporting systems of Crustacea from various habitats (from Shaw[486]).

	Usual medium	Concentration of medium required to half saturate the transporting sites
Carcinus maenas	Sea water and brackish water	20 mE/l Na
Gammarus duebeni	Brackish and fresh water	1·5 mE/l Na
Austropotamobius pallipes	Fresh water (hard)	0·2–0·3 mE/l Na
Gammarus pulex	Fresh water	0·15 mE/l Na

The concentration of sodium in the medium required to give a half maximal rate of transport is the same, whether the anion present is chloride or sulphate. This suggests that the limitation of sodium transport is in no way due to the drag imposed by the anion, but is due to a direct influence of the concentration of sodium on the transporting sites.[484] The concentration of chloride in the medium no longer restricts chloride uptake by crayfish when its level exceeds 0·2–0·6 mE/l, indicating that the level for other ions is not the same as that for sodium.[483] This fact must always be borne in mind in any attempt to relate the distribution of animals to the ionic concentration of their media. A possible way in which the rate of loss of ions, the concentration of the blood and the rate of active uptake of ions may interact is illustrated in fig. 11.

The 'normal' blood concentration is at A, the level at which the rate of uptake with sites fully saturated cuts the normal rate of loss. If the medium is only of a concentration sufficient to half-saturate the sites the blood concentration drops to B, the point at which the rate of loss and the rate of uptake at half-saturation are equal. Similarly, if the loss rate is increased, as might occur at moult, then the blood concentration will fall until loss and uptake are equal and opposite. In each case it will be noted that the change in blood concentration required to effect a new balance point is comparatively small. Similarly only a small rise in blood concentration will occur if the concentration of the medium is raised to half the normal blood concentration (Point C).

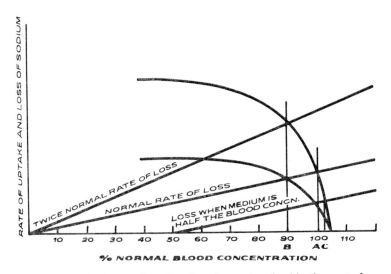

FIG. 11.   The interaction of various factors involved in the control of blood concentration.   (See text for explanation.)

The Effect of Temperature on Ionic Balance

Temperature changes have little effect on the rate of sodium loss by the fresh water isopod *Asellus aquaticus* but they do alter the rates of active uptake.[338]   Thus an increase or decrease in temperature results in an imbalance between the loss and uptake of ions.   A fall in temperature is followed by a drop in blood concentration.   However, the blood comes to a new steady state concentration only a little below the original level because, as already described, the drop in blood concentration activates the uptake mechanism.[338]   A drop in temperature also results in a net loss of sodium by the fresh-water crab *Potamon niloticus*,[479] but not all animals show the same effect.   In *Gammarus duebeni* the temperature coefficient for loss of sodium is larger than that of the active transport system, and so in this animal there is a tendency for the total sodium in the animal to rise after a lowering of the temperature.[296, 336]   Again, however, the change in blood concentration is small, as the blood comes into a new steady state after a small rise.   Additional special compensatory mechanisms apparently come into play during cold seasons of the year.   For example, winter blood concentrations tend to be higher than those of summer in *Asellus*, though the reverse might have been expected.[338]   In *Hemigrapsus* little difference is apparent between the ion concentrations at 5°C in winter animals and at 15°C in summer animals.[128]

The Influence of other Factors

Although the aspect principally studied has been the relationship between the transport and concentration of the blood, it is probable that other factors can also influence the rate of transport. In the case of mammals, blood volume as well as blood concentration governs the output of the hormone aldosterone which controls rate of sodium loss from the body.[177]

Experiments on the brackish-water amphipod, *Gammarus duebeni*,[342] indicate that the rate of active uptake of sodium can be greatly increased in some circumstances even when the blood concentration is very high.   Thus individuals transferred from 160% sea water to de-ionized water show, after about two hours, a greater capacity to take up sodium from a dilute test medium than animals kept continuously in 50% sea water.   This increase in the rate of sodium uptake occurs in the test solution despite the fact that the blood concentration of the animals may, at the time, be twice as high as that of the controls in 50% sea water. *Gammarus* placed from 160% sea water into 50% sea water or sucrose osmotically equivalent to 50% sea water show no increase in the rate of sodium uptake on test.   The full implications of these observations remain to be elucidated, but taken at their face value they seem to suggest that in this animal adjustments can be made to the rate of active transport independently of the concentration of the blood at the time.

Reduction of the blood volume by withdrawing a quantity of blood also brings about an increase in the rate of active uptake of sodium by *Gammarus*.[336]   In this case there cannot be any question that the activation is stimulated by dilution of the blood,[336] and hence it appears that, as in the mammals, volume regulation may be linked to sodium transport.

Site of the Active Uptake of Ions

In forms, such as the frog, which have an unprotected body surface the uptake of ions can occur across any part of the skin. In Crustacea, as in fish, the site of active uptake is limited to specialized regions. Where gills are present, these are the regions primarily responsible for the transport of ions, though not necessarily all the gills participate. In *Artemia salina* the last pair of gills does not stain with silver solutions[116] and is presumed not to play a part in ion transport. Similar non-staining posterior gills have been found in the fresh-water crab *Potamon*.[173] *Asellus aquaticus*, in addition to staining over certain regions of the gills, also has a silver staining plaque on the ventral portion of the abdomen.[264] As the capacity to stain with silver does not prove that the region responsible transports ions, there can be no more than a supposition that ions are taken up by these non-gill regions of *Aselles*. It seems more certain that the silver staining plaques at the lateral ventral corners of the cephalothorax of *Neomysis integer* transport ions, since this animal lacks specific gills.[428]

EXCRETION

The principal excretory organs of the higher Crustacea are formed by a pair of glandular coelomoducts which are situated in either the antennary or maxillary segments. Each gland consists of an end sac representing a coelomic remnant, and excretory tubule opening at the base of the maxilla or second antenna. In most forms the tubule is simple, but in the decapods it may be expanded with much folding of the wall to form a labyrinth at the proximal end and expanded distally to form a bladder (fig. 12*A*).

The principal problems associated with excretion include (1) the manner in which the urine is formed, (2) the function of the urine, (3) the mechanism by which urine composition is controlled.

Mode of Urine Formation

No structure comparable with the glomerulus of vertebrates has yet been reported in the crustacean excretory organ and there

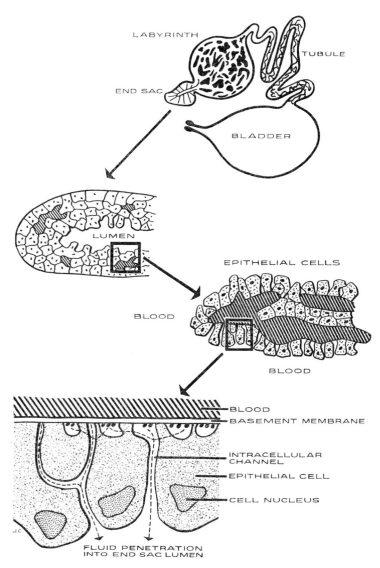

FIG. 12. Diagram of renal organ and the fine structure of the coelomic end-sac, showing a potential site of ultrafiltration. (Redrawn from Kummel [315])

has consequently been considerable controversy as to whether urine is formed by filtration, as in the majority of vertebrates, or by some form of secretory process as in the insect malpighian tubules and kidneys of aglomerula fish.

Martin[363, 364] has now pointed out that some of the earlier physiological results[357, 358] cited as indicating that the antennary gland forms urine by secretion may equally be interpreted as indicating the presence of filtration. More recent work[444, 445, 448] also indicates that a form of filtration occurs.

Evidence for Ultra-Filtration

There are two lines of evidence to support the conclusion that in the decapods urine formation is by a process analogous to the ultra-filtration and re-absorption mechanisms of vertebrates :

(1) Glucose appears in the urine after injection of phlorizin into the blood.[72, 448] (2) The carbohydrate inulin is concentrated in the urine.[72, 448, 449]

Phlorizin inhibits the re-absorption of glucose by the vertebrate nephron. Hence, as glucose appears in the definitive urine of crayfish[448] and Homarus[72] after injection with this drug, there must be a strong presumption that glucose is normally present in the primary urine. Its presence here can be satisfactorily explained only if some form of filtration process is responsible for primary urine formation as it is unlikely to be secreted into the tubule only to be re-absorbed subsequently. Inulin injected into the blood appears in the urine at the same or somewhat higher concentration in Homarus,[72] Carcinus maenas,[449] Hemigrapsus nudus,[447] Ocypode albicans,[187] Austropotamobius pallipes,[444] Procambarus clarkii, Orconectes virilis and Pacifastacus leniusculus.[448] Since the lobster, like the vertebrates, is not able to metabolize this substance,[199] its presence in the urine at a concentration similar to or higher than that of the blood suggests urine formation by filtration with or without subsequent re-absorption of water.

If filtration by hydrostatic pressure is to occur, the blood pressure must exceed the sum of the hydrostatic pressure in the urinary tubule and the colloid osmotic pressure of the blood proteins. As the crustacean excretory organs are bathed by blood it may be expected that the hydrostatic pressure within the organ

is the same as that in the surrounding sinus.   Filtration is therefore unlikely to occur from the general haemocoel against the
colloid osmotic pressure.   However, all parts of the antennal
gland of the crayfish[354, 359, 412, 448] and the end sac of *Palaemonetes*[404]
receive branches of the antennary artery, in which the pressure
must be higher than that in the haemocoel, since there is a through
flow of blood.

In the coelomosac there is a histological arrangement which is
compatible with ultra-filtration, though—as already stated—not
resembling the vertebrate glomerulus.   Separating the blood
from the coelomosac lumen is a cellular epithelium lining a thin
basement membrane.   The epithelial layer does not necessarily
constitute a cellular barrier, since the cells tend both to be somewhat separated from one another and also scalloped at their bases
(fig. 12*D*), so that there is potentially a free passage to the coelomosac lumen for material able to traverse the basement membrane.[315]

This anatomical evidence that the coelomosac is a possible site
of ultra-filtration is supported by physiological experiments.
Following injection of inulin into the blood of crayfish it appears in
the coelomosac in almost the same concentration.[445]   The same
is true of dextrans with molecular weights of 15,000 to 20,000.[296b]
The fact that the concentration of these non-metabolizable substances is the same in the coelomosac as in the blood implies that
they have been filtered from the blood.   The size of molecule
filtered by the crayfish antennary gland is somewhat higher than
the vertebrate limit, which lies in the range 37,800 – 62,000.   The
upper limit for the crayfish appears to be around 150,000 though
there is some limitation of filtration rate even with dextrans with
molecular weights of 60,000 to 90,000.[296b]

Unfortunately such few studies as have yet been made of the
pressure in the arteries and sinuses do not give an unambiguous
answer as to whether or not the pressure difference between
the arterial and sinus bloods is large enough to exceed the colloid
osmotic pressure and bring about filtration.   The intraventricular
pressure of the lobster, *Homarus americanus*, is about 9 mm of
mercury in excess of the sinus pressure at the base of the antenna[73]
and thus might conceivably be sufficient to ensure filtration.   In
Cancer and *Maia* the corresponding values are 8 mm[129] and 4 mm[148]

respectively. The colloid osmotic pressure of these animals has not been determined; but if the pressure values quoted are correct there can be but little margin to ensure filtration. It must be remembered, however, that the open blood system of Crustacea makes the recording of differences between arterial and sinus pressures difficult, since any opening of the system is liable to distort the values. If it is subsequently shown that the hydrostatic pressure is indeed inadequate to account for the movement of fluid into the excretory organ in all cases, other systems compatible with the observed presence of glucose and inulin will have to be considered. One possible mechanism might be that excretion of an ion or ions creates an osmotic or electrical gradient across the wall of the excretory tubule and this in turn results in a bulk flow of fluid into the tubule by osmosis or endosmosis. Some such system would seem to be necessary in those smaller Crustacea which lack arteries or even a heart, though conceivably the hydrostatic pressure in the surrounding sinuses could cause pressure filtration if the wall of the excretory organ were so supported by connective tissue that the tubule pressure was less than that of the blood.

Apart from filtration and the alternatives suggested above, there is also another means available by which materials could be excreted via the antennary gland. Studies on the primary urine of crayfish have shown that formed bodies are present in the lumen of the coelomosac, labyrinth and excretory tubule. These bodies appear to be extruded portions of the epithelial cells. They form a possible alternative to filtration for the removal of large molecules from the body. For example, the dye Congo red, which accumulates selectively in the cells of the coelomosac after injection into the blood, subsequently appears in the formed bodies.[446c] Similarly, injected foreign protein labelled with fluorescent dye also accumulates initially in the coelomosac cells.[286b] Pinocytosis vesicles are formed at the base of the coelomosac cells [315] and these therefore present a possible route by which large molecules to be excreted could be taken up by the coelomosac cells from the blood. The formed bodies contain proteolytic enzymes which may be presumed to attack proteins excreted in this way. Since the formed bodies disrupt before they reach the bladder [446c] and since the bladder wall is known to be an area of, at least, ion-transport [284] it is not improbable that useful products of degraded excretory material can subsequently be re-absorbed.

Functions of the Excretory Organ

The excretory organ has several functions the relative importance of which varies with the habitat in which the species lives. These functions include (1) elimination of metabolic waste materials too large to diffuse across the cuticle, (2) regulation of the ionic composition of the blood, and (3) regulation of the body volume. In cases where the blood is hypertonic to the medium, the urine of a few forms may also assist in the maintenance of this hypertonicity.

(1) *Excretion of waste materials.* An excretory organ forming urine by filtration offers a convenient means by which foreign or waste materials, e.g. inulin, for which there may be no available specific transport mechanism, can be readily removed from the blood. Some organic substances can be actively secreted into the tubule however. Thus, if phenol red is injected into a lobster the concentration of the dye in the urine exceeds the concentration of inulin, thus indicating that the former is not concentrated solely by the withdrawal of water passing down the urinary tubule.

Some 60–87%[129, 145] of the total nitrogenous waste material is in the form of ammonia in marine and fresh-water forms. Ammonia, being a toxic material, requires a considerable volume of water for its elimination, and terrestrial groups, such as the gastropods, reptiles, mammals and birds, tend to release waste nitrogen in the form of the less toxic urea or uric acid, to conserve water. There is little evidence that such a switch occurs in terrestrial Crustacea, ammonia still accounting for some 45–55% of the total nitrogen output in the terrestrial isopods. Urea rarely accounts for more than about 5% of the total nitrogen loss, though 13 and 12% have been reported respectively in *Cancer pagurus*[129] and *Astacus astacus*.[129] Uric acid accounts for up to 8% in the isopods [145] but is absent, or occurs only in small amounts, in other forms. Indeed, instead of combining waste nitrogen in a non-toxic form the terrestrial isopods tend to reduce the total nitrogen turnover.[145] Amino-nitrogen is often present in the excretory products, constituting usually between 5 and 20% of the total, though these substances are absent in *Carcinus*.[178]

Amino-acids are released by *Calanus* when the animals are kept in overcrowded conditions (15 per cc.) but not when they are given more room.[111b]   It is possible therefore that the release of amino nitrogen may be a response to unfavourable conditions and that these substances are not normal excretory products.

There is little evidence to show by what routes these various end products leave the body.   Ammonia and urea can diffuse across the gills of the lobster[72] and presumably of other aquatic forms.   Up to 30% of the ammonia is said to be lost as a gas from wood lice.[145]   The urine is likely to be used as a vehicle for larger, less readily diffusible, waste products.   Quite large molecules can be filtered by the antennary gland of the lobster, the urine turning red after injection of dog-fish haemoglobin.[72]

(2) *Ion regulation*.   It has already been noted that the ionic composition of the blood of marine species differs from that of the medium.   The very fact that urine is formed at all would contribute to a decrease in the blood concentration of those ions which can only slowly penetrate across the cuticle, even if the urine were similar in composition to the blood.[397]   The excretory organ does not, however, play only a passive role in ionic regulation, since the ion levels in the urine differ from those in the blood (table 8).   Ions such as magnesium and sulphate, which usually have a lower level in the blood than in the medium, tend to have a high concentration in the urine.[429]   Furthermore, the ion ratios in the urine are often altered when the concentration of the medium is changed.   This active role of the antennary gland in controlling blood ion levels can be illustrated by the way in which the magnesium ion is handled.   Increases in the magnesium concentration of the medium are followed by an increase in the magnesium concentration of the urine but little change in blood concentration in *Carcinus*,[539] *Hemigrapsus*[128] and *Cancer*.[232]   At first sight, the shore crab *Pachygrapsus crassipes* seems to behave in the same manner, since the ratio of urine magnesium to blood magnesium rises from 5·6 when the animal is in 50% sea water to 15·4 when it is in 150% sea water.[230]   However, determination of the amount of magnesium lost in 50%, 100% and 150% sea water indicates that the animal actually loses most magnesium in the dilute

medium and least in the most concentrated, because of differences in the urine volume.[235]   This finding, of course, drives home the importance of determining urine volume as well as concentration, before conclusions are drawn as to the function of the excretory organ in regulation.   It appears therefore that the concentration of magnesium in the urine of *Pachygrapsus* is largely determined

TABLE 8

The ionic composition of the blood and urine of Carcinus maenas *in sea water and after 48 hr and 96 hr in water-saturated air* (from Riegel and Lockwood)[449].

	In sea-water	48 hr	96 hr
Na mM/l			
Urine	481 ±37·8	459 ±30·8	391 ±50
Blood	525 ±10·9	538 ±18·9	543 ±19·5
K mM/l			
Urine	12·4 ±0·57	12·1 ±0·57	13·4 ±0·79
Blood	12·7 ±0·85	12·4 ±1·39	13·6 ±0·66
Ca mM/l			
Urine	17·2 ±1·25	21·3 ±2·24	22·3 ±5·03
Blood	14·3 ±1·41	14·6 ±1·10	15·2 ±2·21
Mg mM/l			
Urine	89·4 ±35	121 ±47	192 ±37·0
Blood	21·2 ±2·9	20 ±2	19·4 ±3·3
Cl mM/l			
Urine	531 ±35·6	671 ±86	626 ±14·6
Blood	502 ±22·5	569 ±26·8	558 ±19·5
Inulin (% original)			
Urine	202 ±27	187 ±43	211 ±30
Blood	100 ±0	81 ±10·5	87 ±6·6
Osmotic pressure≡mM/l NaCl			
Urine	559 ±4	—	646 ±19
Blood	568 ±14	—	622 ±15

by the volume of urine produced rather than by the requirements of the maintenance of the blood magnesium level.   As the concentration of magnesium in the urine rises, the ratio urine concentration to blood (U/B) for sodium declines in the crabs *Pachygrapsus*,[232, 427] *Carcinus*,[449] *Ocypode*,[232] *Uca*,[232] and *Hemigrapsus*.[232] This must mean that the excess sodium is excreted elsewhere. The actual site of secretion is the gill in the case of *Pachygrapsus* and presumably also in the other forms too.

Gecarcinus, one of the most fully terrestrial of the land crabs, is a form which visits the sea only for reproduction, and hence does not normally have to dispose of excess magnesium. Interestingly this animal does not possess the ability to concentrate magnesium in the urine even when magnesium is injected into the blood.[231, 232]

Some doubts surround the mechanism whereby the U/B of magnesium is raised above unity in the crabs listed above. The concentrating of magnesium in the urine might be brought about either by an active secretion of magnesium into the tubule or by withdrawal of water plus sodium. Investigation of the U/B for inulin suggests that about half the primary urine initially formed by *Carcinus* is re-absorbed during passage along the urinary tubule.[449] As the definitive urine remains isotonic with the blood, salt must also be re-absorbed. This re-absorption will serve to concentrate in the tubule substances which are not taken back; but it cannot be solely responsible for the differences between the magnesium concentration of the urine and blood, suggesting that magnesium can be secreted into the tubule.[449]

Species in which the blood concentration is maintained hypertonic or hypotonic to the medium tend to lose and gain salts respectively. The production of urine hypotonic to the blood by hypertonic forms, and of urine hypertonic to the blood by hypotonic forms, would therefore be a valuable contribution towards the maintenance of the gradient between blood and medium. Nevertheless the production of urine differing in osmotic concentration from the blood as an aid to hyper- and hypotonic regulation is not common in the larger Crustacea. All the marine and brackish-water decapods seem to produce only isotonic urine. So too do the fresh-water crabs and prawns such as *Eriocheir sinensis*,[468] *Potamon niloticus*[479] and *Palaemonetes antennarius*.[405] On the other hand the brackish-water amphipods *Gammarus duebeni*,[339] *G. zaddachi*[339] *G. fasciatus*[548] and the fresh-water crayfishes *Procambarus clarkii*,[334] *Austropotamobius pallipes*, *Astacus* sp.,[468] *Orconectes virilis* and *Pacifastacus leniusculus*[448] and the amphipod *G. pulex*[335] all have the capacity to produce urine hypotonic to the blood. Three genera, *Ocypode*,[187] *Goniopsis*[187] and *Uca*[225] have been claimed to form urine whose ion concentration is either greater or less than that of the blood, but re-investigation on *Uca* and *Ocypode*[232] has failed to confirm that the difference between blood and urine concentration is sufficiently great for the latter to be claimed as playing any major role in the hypotonic regulation of the blood.   The question whether or not Crustacea are able to form urine more concentrated than the blood must therefore remain unanswered for the time being.   The more terrestrial of the land crabs such as *Gecarcinus lateralis*,[231] *Coenobita brevimanus*[233] and *Birgus latro*[233] do not form hypertonic urine.

The production of urine hypotonic to the blood eliminates excessive water and helps to conserve ions in the body.

The capacity to produce hypotonic urine can be related to the structure of the excretory organ.   In decapods which form isotonic urine there is only a short duct separating the labyrinth from the bladder, whilst in crayfish there is a long duct.[412] Similarly in gammarids the marine *G. locusta* has a short excretory tubule whilst that of the fresh-water *G. pulex* is more than twice as long.[274, 471]   *G. duebeni*, which has a somewhat shorter tubule than that of *G. pulex*, can form hypotonic urine; but the minimum concentration (*circa* 40 mM/l) produced is not so low as that of *G. pulex* (*circa* 5 mM/l).[339]

In the crayfish the primary urine in the end sac and proximal parts of the labyrinth is practically isotonic with the blood, but the urine in the bladder is only a little more concentrated than the fresh-water medium.   Dilution must therefore occur in the intervening tubule.[412, 445] As the urine in the distal tubule is considerably more concentrated than that in the bladder of the crayfish it must be presumed that the bladder wall also plays a part in the further dilution of the urine[445] (table 9).   The role of the bladder wall in re-absorbing ions is further evidenced by the presence in the bladder wall of cholinesterase, a high concentration of which is often found in cells associated with sodium transport.   Furthermore, inhibition of this substance with eserine partially blocks the normal movement of ^{22}Na from the bladder into the blood.[284]

(3) *Regulation of body volume.* Some compensation for increased blood volume may be possible by decreasing the volume of the bladder, but the possibility of expansion within a rigid exoskeleton is limited.   The volume of Crustacea must therefore be regulated within narrow limits if fluctuations of the internal hydrostatic pressure are to be avoided.   That an effective mechanism

is involved in volume regulation is clearly indicated by the fact that the size increase in crabs (*Callinectes*) is constant at moult, no matter what the gradient of salinity between their blood and medium.[257]

Crabs respond to a sudden loss of blood volume by drinking the medium. Thus when blood is removed from *Carcinus* they can regain much of the weight lost within a few minutes of transfer back to sea water.[365b]

TABLE 9

The osmotic pressure and chloride concentration in blood and in the different regions of the crayfish antennal gland (from Riegel[445]).

	Blood	Coelo-mosac	Laby-rinth	Tubule proximal	Tubule distal	Bladder
O.P. mean ±S.D.	200 ± 10	208 ± 22	181 ± 11	169 ± 7	121 ± 14	$18 \cdot 8 \pm 1 \cdot 4$
Cl mean ±S.D.	184 ± 24	168 ± 20	162 ± 21	132 ± 24	$80 \cdot 8 \pm 15$	$3 \cdot 37 \pm 1 \cdot 8$

The excretory organs play an obvious role in regulating the fluid volume. Control of the rate at which water is eliminated is achieved either by regulation of the rate at which primary urine is formed or by re-absorbing fluid during passage of urine along the excretory tubule. Both systems are used, but the forms specialized for water retention in the body seem to concentrate on the latter. When the lobster *Homarus* is injected with inulin, it forms urine containing the same concentration of this substance as the blood, i.e. $U/B = 1$[72], suggesting that there is no subsequent water withdrawal from the tubule. The animal's ability to vary the rate of filtration is indicated, however, by the finding that the rate of clearance of inulin from the blood is variable.[72] An extreme example of control of filtration rate is shown by *Ocypode albicans*. This semi-terrestrial crab varies the rate of inulin clearance (filtration rate) according to the salinity of the medium in which it is placed; when taken out of water it ceases to clear inulin at all.[187] Failure to clear inulin implies either that filtration has ceased or that there is no re-absorption of water from the urine into the blood.   Cessation of filtration might be an adequate mechanism in a form which has the opportunity to produce urine during its nightly visits to the sea.   In a more fully terrestrial form, such as *Gecarcinus* which does not visit the sea daily, cessation of primary urine production would be expected to involve an unacceptable retention of metabolic waste products. When on land, *Gecarcinus* continues to clear inulin, effecting a clearance of a volume equivalent to the blood volume in sixty hours.   During this time, however, no definitive urine could be collected by Flemister,[187] suggesting that most of the primary urine had been re-absorbed.   Naturally when the animals are deprived of water complete re-absorption is necessary to avoid diminution of the blood volume.   Re-absorption of water by *Gecarcinus* seems to be entirely directed at concentration of waste production and regulation of fluid volume rather than of blood concentration; for it will be recalled that this species does not form urine hypertonic to the blood.   Some re-absorption of water also apparently occurs from the excretory tubules of the crayfishes *Procambarus clarkii*, *Orconectes virilis* and *Pacifastacus leniusculus* judging from the fact that the U/B for inulin exceeds 1. It seems likely, however, that this represents an obligatory loss of water down the osmotic gradient produced during dilution of the urine rather than an active withdrawal of water under facultative control.

Urine Volume

The classical method of determining urine volume was to block the excretory pores and then determine the rise in body weight over a period of time.   This method is open to the obvious criticism that increasing body weight may result in a hydrostatic pressure increase which impedes normal water entry and so gives a low value for urine production.   It has been found, however, that the results for the crayfish do correspond fairly closely to those of inulin clearance studies.[444]

The volume of urine produced varies according to the osmotic gradient between blood and medium, the permeability of the body surface, and the size of the animals.   It may also vary in individuals of the same species, the volume decreasing with a fall in temperature,[444] a lowered blood colloid osmotic pressure,[72] or as a result of shock during handling.[448]   As a consequence of such factors and differences in the methods of determination, the values quoted in the literature range rather widely.   The marine

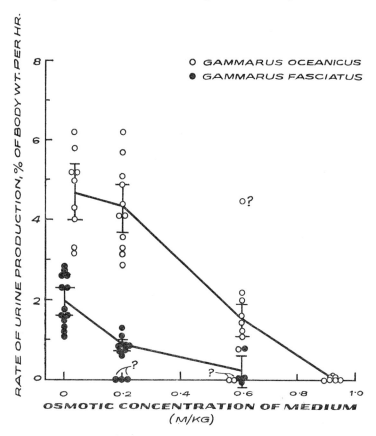

Fig. 13.   The relation of the rate of urine flow to the external concentration in *Gammarus oceanicus* (marine) and *G. fasciatus* (freshwater).   (From Werntz.[548])

decapods seem in general to have a daily urine output equivalent to about 3–15% of the body weight, *Carcinus* 3·6%,[487] 4·7%,[539] 10%,[381] *Cancer* 3–10%[454] *Homarus, Pachygrapsus* 3·9%[235]. Responses to media more concentrated than the blood vary.   Both

Pachygrapsus crassipes and *Palaemonetes varians* maintain the blood hypotonic to concentrated media; but, whereas in the former the urine volume is lower in 150% sea water than in 100% sea water,[235] the latter has its lowest urine output when the medium is isotonic with the blood and increases the output in more concentrated media.[404] Possibly this rise is associated with the need to eliminate excess magnesium or sulphate entering by diffusion.

The urine flow-rate is increased when there is an osmotic gradient drawing water into the blood. Thus, in 50% sea water *Pachygrapsus* is estimated to produce a urine volume equivalent

TABLE 10

	Approx. wt.	Blood concentration mM/l NaCl	Approx. urine vol./day, as % body wt.
Daphnia magna	10 mg	60[211]	**200**[513]
Gammarus pulex	60 mg	150[339]	**47**[339]
Asellus aquaticus	60 mg	150[337]	**25**[337]
Procambarus clarkii	30 gm	183[334]	5·3[334]
Austropotamobius pallipes	30 gm	227	8·2[67]
Potamon niloticus	20 gm	271[479]	0·6–0·05[479]

Numbers in heavy print are urine volume as percent total body water daily.

to 58% of its body-weight daily,[235] nearly fifteen times the rate in sea water; and *Palaemonetes varians* in 5% sea water has a urine volume of 39% of the body-weight a day as opposed to 3·6% in 50% sea water.[404] Detailed studies of the amphipods *Gammarus oceanicus* and *G. fasciatus* indicate that urine output is linearly related to the osmotic gradient (fig. 13).[548]

Fresh-water species vary widely in their rate of water turnover. Very impermeable species such as *Potamon niloticus* produce even less urine than marine forms, despite the large osmotic gradient between blood and medium. Species of smaller adult body size, because of their larger surface to volume ratio, would be expected to have high rates of water turnover; and this is indeed the case despite the fact that lowered blood concentration of most smaller forms decreases the osmotic gradient (table 10).

It has been suggested[340] that the low blood concentration of the smaller fresh- and brackish-water species can be correlated with the need to minimize the osmotic work.   For species with similar surface permeabilities to water and ions, the smaller the size the greater will be the turnover of ions.   The larger the turnover of ions the greater will be the energy expenditure *per unit mass of body tissue* in order to maintain a given gradient of concentration between blood and medium.

Control of the Rate of Urine Output and Concentration

Practically nothing is known of the way in which the urine volume is regulated, though the fact that crayfish may become anuric when handled roughly[448] and that urine production continues normally after withdrawal of 10% of the blood volume of the lobster[72] suggests that more than purely passive processes are involved.   The U/B for inulin declines when crayfish are eliminating an excess water load, indicating that less water than usual is being re-absorbed[444]; but this might be the result of a faster flow of urine through the excretory tubule allowing less time for passive re-absorption rather than an active effect.   Similarly, little is known of the mechanisms controlling the concentration of the urine.   The fresh-water amphipod, *Gammarus pulex* seems to have little facultative control over urine concentration, since it continues to produce urine very hypotonic to the blood even when it is in a medium more concentrated than the urine.[339]   *Austro-potamobius* behaves rather similarly, though the urine concentration does increase somewhat and tends towards isotonicity with the medium at the highest salinities the crayfish will tolerate.[65]   The brackish-water amphipods *G. fasciatus*, *G. oceanicus*[548] and *G. duebeni*[339] by contrast maintain the urine more concentrated than the medium though less concentrated than the blood when they are in diluted sea water (fig. 14).   The concentration of the urine of *G. duebeni* can be correlated with the concentration of the blood when the animals are in a steady state.[339]   Blood concentration, however, is not the only factor influencing urine concentration; transference from 110–175% sea water to fresh water results in the animals forming urine hypotonic to the blood even when the concentration of the blood is twice as high as the level at which hypotonic urine would be expected to be produced by animals in a steady state with their medium.[339] As *G. duebeni* does not produce hypotonic urine when transferred from 175%

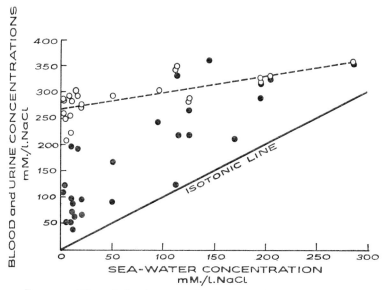

FIG. 14.   The relation between the concentration of blood, urine and medium in *Gammarus duebeni*. ○ Blood concentration, ● Urine concentration. (From Lockwood.[339])

sea water to 50% sea water (in which medium the urine of steady-state animals is isotonic with the blood), the tentative conclusion has been drawn that this species may be able to monitor the concentration of its medium and make appropriate adjustments to its urine concentration on this basis.

4. Terrestrial Crustacea

All the terrestrial Crustacea are members of one of three orders, the Decapoda, Amphipoda and Isopoda. Other, smaller forms, mainly Ostracoda and Copepoda, but including *Bathynella* of the Syncarida, occur on land; but, as they are confined to the interstitial soil water,[386] wet moss and similar places where there is a film of water, they cannot be regarded as truly terrestrial.

Most of the animals now living on land, including the terrestrial Protozoa, Turbellaria, Annelida and vertebrates, are thought to have been derived from ancestors which lived in fresh water. Many of the land gastropods too may have evolved from fresh-water or brackish-water forms. Terrestrial Crustacea on the other hand seem more likely to have invaded the land directly across the marine littoral strip with perhaps, in come cases, ancestors which lived in land-locked saline lagoons.[269] Of the three orders the decapods are the least emancipated from the sea. Most, or perhaps all, land crabs have larval stages and so, since at least the females have to return to the sea to release their off-spring, the terrestrial range of crabs is limited to a few hundreds of yards from the shore. The amphipods and isopods, retaining the eggs within a brood pouch and having compressed the develop-mental stages so that the young are released as immature tiny adults, no longer have an aquatic stage in the life cycle; so they are not limited in their range by the necessity of remaining close to a large body of water. Both groups colonize land widely though the former are limited, as fully terrestrial forms, to the Indo-Pacific region.[269]

As far as regulation of the body fluids is concerned, the terres-trial habitat exposes animals to the two hazards of overhydration and desiccation. Temporary flooding of the ground by rain is unlikely to present an osmotic problem to animals derived from fresh-water ancestors, since most will have inherited from these ancestors the capacity to eliminate excess water in the form of hypotonic urine. However, as all the marine Crustacea so far studied produce urine only isotonic or nearly isotonic with the blood, the terrestrial Crustacea are unlikely to have inherited the facility to dispose of excess water without an accompanying salt loss. Fresh-water crabs, though continuously exposed to osmotic water uptake, nevertheless produce urine isotonic with the blood and replace the lost salt by the active uptake of ions from the med-ium. Salt cannot be so readily replaced on land, as it is usually available only in the food, but despite this none of the terrestrial crabs seems to have developed the capacity to produce dilute urine. The urine concentration of land isopods and amphipods has not yet been studied.

Crustacea lack the epicuticular wax layer which so effectively decreases water loss across the body surface of insects[28, 166] and consequently lose water by evaporation when exposed to unsaturated air. The rate of water loss (E) is potentially equal to the product of the saturation deficit of the air $(S.D.)$ and the permeability (K) of the cuticle to water $(E = K \times S.D.)$. Saturation deficit is the vapour pressure of water in saturated air less the actual vapour pressure, $(S.D. = V(100 - RH))$ where V is the vapour pressure of saturated air at a given temperature and RH is the relative humidity. As the saturation deficit is a measure of the drying capacity of the air, under ideal conditions the rate of water loss should be proportional to it. In practice it is difficult to predict with accuracy the rate of water loss even from inert materials for two reasons: (1) Gradients of humidity build up near the surface and tend to slow the evaporative loss. (2) The cooling effect of evaporation alters the vapour pressure of the liquid at the surface. In the case of evaporation from animals the rate of water loss is even less predictable, since the permeability of the surface itself may be affected by both degree of hydration and temperature.

Despite these factors tending to diminish evaporative water loss, it is nevertheless sufficiently rapid to constitute a problem. Hence it is important that terrestrial Crustacea either limit their permeability or remain in areas where the humidity is high. Limitations are placed on the restriction of evaporative losses by the necessity of keeping the respiratory surfaces moist so as to ensure adequate gaseous exchange. Consequently, the avoidance of evaporative water loss largely depends on behavioural mechanisms which serve to keep the animals in areas of high humidity.

In addition to evaporation, water is also lost in the removal of nitrogenous waste. In aquatic forms the end products of nitrogen metabolism may be readily eliminated across the gills as NH_3 or in exchange for sodium ions as NH_4^+. Only small amounts of ammonia can be released across the surface in most land forms (though up to 30% of the total in some isopods) and most of the remainder probably passes out in the urine, with a consequent loss of water. In the terrestrial insects and arachnids the nitrogenous waste is concentrated as insoluble products, uric acid and guanine, which can be eliminated as a sludge with a minimal waste of water. Ammonia, however, remains as the

5

main nitrogenous waste product in the Crustacea, at least in the terrestrial isopods and *Orchestia*.[145]   Uric acid accounts for only some 5–10% of the excreted nitrogen.   There is thus little tendency towards the modification of the metabolism to produce relatively non-toxic end-products as in vertebrates.   The principal metabolic concession towards the need for water conservation by the isopods seems indeed to be that there is a general depression of nitrogen metabolism.[145]   In woodlice some of the uric acid that is formed is retained and stored in the body.   Forms such as *Porcellio* and *Armadillidium* contain more stored uric acid than species which are less well able to tolerate short term exposure to dry air. However, as the fresh-water isopod *Asellus aquaticus* contains considerably more stored uric acid than *Armadillidium*,[145, 270, 341] it seems unlikely that storage of this material represents simply a response to lack of an adequate supply of water for excretion.

In special circumstances the form of nitrogenous excretion may vary.   Thus it has been found that when *Calanus* are kept in overcrowded conditions (15 animals per cc) they release amino acids, though these substances are not normally excreted.[111b]

TERRESTRIAL DECAPODA

The Anomura and Brachyura contain a series of species showing different degrees of independence of the sea and physiological and behavioural adaptations to life on land.   The series is of interest in that it may provide some insight into the stages of physiological evolution through which the more fully terrestrial species have passed.   The following species will be considered, *Emerita talpoida*, *Pachygrapsus crassipes*, *Uca* sp., *Ocypode albicans*, *Coenobita perlatus*, *Coenobita brevimanus*, *Birgus latro* and *Gecarcinus lateralis*.

Emerita talpoida

The mole-crab lives in the wave-stirred zone of sandy beaches, moving up and down with the tide and feeding in the film of water left by each receding wave.   It swims and burrows readily and is not normally exposed to the drying action of air.   In common with the sub-littoral crabs, it possesses no capacity to regulate the concentration of its blood though it does tolerate a limited degree of concentration change since it can survive in both 75% and 125% sea water for twenty-four hours.[227]   Exposure to concentrations of 150% or 50% sea water proves fatal in a few hours.[227]

Pachygrapsus crassipes

This is a crab of the mid- to upper-tide level of rocky shores. The blood is close to isotonicity with normal sea water but, unlike that of *Emerita*, is regulated when the animal is in more concentrated or dilute sea water, such as it might occasionally meet in rock pools.   Thus in 160% sea water the blood concentration is osmotically equivalent to 140% sea water and in 25% sea water to about 70% sea water.[227]   The urine of *Pachygrapsus* plays little part in this regulation of the blood concentration since it is usually isotonic with the blood[227, 281, 427]; but urine volume is presumably accurately controlled to meet the volume regulation of the animal, as there is no measurable weight change on transference between these media.[227]   On desiccation to a 7% weight loss there is a shift of potassium, calcium and magnesium from the tissues into the blood, a physiological failure which may limit the terrestrial life of the species.[230]   In its behaviour, too, *Pachygrapsus* is not well adapted to life on land; for, if given a choice of 50%, 100% and 150% sea water after concentration of its blood by desiccation, it usually selects 100% sea water even if this is not the medium which would most rapidly restore the blood concentration to normal.   As the animal will normally spend about twelve hours a day in water[228] this preference for sea water would seem to be a powerful factor tending to restrict this species to the inter-tidal and sub-tidal zones of the sea.[228]

Uca

Three species of fiddler crab *U. pugnax*, *U. pugilator* and *U. minax* show, in that order, decreasing dependence on the sea and greater tolerance of fresh water.   *U. pugnax*, a species restricted to the upper shore and tidal salt marsh, has a LD 50 (time for 50% mortality of the population) of only one-and-a-half days in fresh water, *U. pugnax* from *Salicornia* marsh, where it may be exposed to rain storms, has a LD 50 of three-and-a-half days in fresh water, whilst the brackish-water form *U. minax*, which often invades streams, has one of three weeks.   When given a choice of media, *U. pugnax* and *U. pugilator* tend to select sea water like *Pachygrapsus* but *U. minax* selects fresh water.[502]   Nevertheless, many of the pools on the marshes where the animals occur will frequently contain water more saline than sea water and appropriately the species of *Uca* have a well developed capacity to maintain a blood concentration hypotonic to strongly saline media.   Indeed the concentration of the blood rises to a level only a little above that of sea water itself when they are kept in 175% sea water for two to four days.[225]

There is little advance over the *Pachygrapsus* state with respect to the urine however.   This cannot play a major role in the maintenance of hypotonicity of the blood, since at most it is only slightly hypertonic to the blood in sea water or more concentrated media.[225]   The sodium concentration of the urine is always lower than that of the blood, indicating that excess sodium is eliminated elsewhere,[225] probably via the gills, as in *Pachygrapsus*.

Ocypode albicans

Ghost crab burrows are usually situated above the high tide mark though they extend down into moist sand.   Unlike the diurnal *Uca* species, the animal remains in its moist and humid burrow during the day and thus limits evaporative water loss. This restriction of water loss is important since, although it visits the sea at night and can there re-hydrate, the blood chloride concentration is maintained at only 380 mM/l[231]; so excessive replacement of water involves considerable osmotic work.

Coenobita

The anomuran *Coenobita* does not regulate its blood concentration within such a narrow range as *Uca* and *Ocypode*.   To compensate for this, the animals show a wide range of tolerance. Thus *Coenobita perlatus* (caught in the wild on Eniwitok Atoll) have been found to have a blood osmotic concentration in the range 102−150% sea water,[233] whilst the limits of tolerance found in the laboratory are from 83−220% sea water.[234]   When given a choice of land, fresh water and sea water the animals spend about 90% of their time on land, a very much greater proportion than was found in the case of *Pachygrapsus*.   Sea water is normally preferred to fresh water for the remaining time, but if the blood concentration is high fresh water will be selected,[234] a further behavioural advance over the situation in *Pachygrapsus*. One factor contributing to *Coenobita*'s ability to remain out of water for such a large part of the day is that it fills the snail shell which it carries, like other hermit crabs, with sea or brackish water. This water comes quickly to osmotic equilibrium with the blood[233] and can therefore be regarded as an extension of the haemolymph as far as its water content is concerned. Another factor is that *C. perlatus* is generally active at night[234] when evaporative water loss is likely to be lower than during the day. *C. perlatus* commonly lives among the tree roots close to the sea, and it makes periodic visits to refill the shell with sea water.

Coenobita brevimanus shows a further advance on the *C. perlatus* condition. This former animal, when given a choice of media to fill its shell, prefers fresh water and is adept at filling the shell from very small puddles; a pool 4 cm long and 1 cm deep suffices.[233] *C. brevimanus* lives further from the sea than its cogenor and is restricted to areas of heavy vegetation, often burrowing in piles of rotting coconuts. By burrowing in wet detritus the animal is able to remain alive for some weeks, even in the absence of free water. Under natural conditions its blood concentration is equivalent to about 80% sea water and this concentration is more uniformly maintained than in the case of *C. perlatus*.[233]

Birgus

Birgus, the coconut crab, though also an anomuran, does not conceal its abdomen in a gastropod shell. Instead, when inactive it remains in moist burrows, thus reducing evaporative losses of water. The blood concentration is usually maintained at a level a little lower than that of *C. brevimanus*, at about 75% sea water. It will select either sea water or fresh water to drink according to need, although it normally seems to show a preference for the latter. Nevertheless, *Birgus* displays considerable ability to tolerate changes in blood concentration and can survive as long as seventy-eight days if provided with sea water alone, even though during this time the blood concentration rises to a level equivalent to 118% sea water.[226] Its osmotic regulation is thus largely based on behavioural control of drinking and the capacity to withstand variations in blood concentration.

Gecarcinus lateralis

Gecarcinus, one of the most terrestrial of the brachyuran crabs, rarely enters water. Like *Birgus* and *Coenobita*, it displays a very considerable toleration of changes in blood concentration. Thus after 35–40 days when given only fresh water the blood sodium concentration had reached 329 mM/l, whilst after 6–8 days with only sea water provided it is about 600 mM/l. When given a choice of both media the blood concentration is maintained at about 460 mM/l.[231] After a period of dehydration *Gecarcinus* can regain its normal weight if allowed to burrow in sand moistened with fresh water, even if no free water is available.[231, 38] It can live up to four months in sand moistened with fresh water, though in sand moistened with sea water there is no weight gain and survival is short. At the posterior, lateral corners of the body there is a system of modified bristles which form a capillary channel leading up to the greatly enlarged pericardial sacs. It is probable that this system is involved in conducting water from moist sand to the sacs where it is absorbed into the body. *Gecarcinus* maintains a more constant blood concentration than does *Pachygrapsus* when desiccated by the same water loss.[231] Another feature in which it shows a better degree of adaptation to desiccation than *Pachygrapsus* is that for a given water loss it shows a much smaller loss of potassium from the tissues.[231]

When *Gecarcinus* is given access only to sea water, its survival is much shorter than either that of *Birgus* or *Coenobita*, indicating the greater removal of the former from physiological adaptations to marine conditions.[231] Probably this is due to a lessened ability to secrete sodium across the gills, a faculty utilized at least by *Pachygrapsus* and *Uca* species of the other crabs mentioned above,[225] but no longer required in a form which rarely has access to salt water. In most semi-terrestrial crabs there is a well developed capacity to concentrate magnesium in the urine, whilst at the same time decreasing the sodium concentration. This capacity has been lost in the more fully terrestrial *Gecarcinus*.

Whilst there is too much variety in the details of osmotic and ionic regulation of these semi-terrestrial and terrestrial forms to make over-all generalizations, certain possible trends are nevertheless apparent. Semi-terrestrial crabs such as *Pachygrapsus, Uca* and *Ocypode*, which always have access to the sea or saline-water pools, have developed a capacity to regulate the blood osmotic and ionic concentration to a considerable degree. The excretory organ is specially adapted to regulate magnesium levels, and excess sodium is secreted across the gills. More fully terrestrial crabs such as *Birgus, Coenobita* and *Gecarcinus*, which do not always live close to the sea, depend more on behavioural mechanisms to regulate the blood concentration and tend to select fresh water to drink. Water is presumably not always available, particularly on dry atolls, and these latter genera have developed a greater tolerance of variations in body fluid concentration than that possessed by the first three genera mentioned. As high intakes of magnesium are uncommon away from the sea, at least some of the more terrestrial forms have lost the capacity to concentrate this ion in the urine.

AMPHIPODA

The terrestrial Amphipoda all belong to the Talitridae which are most familiar to inhabitants of the Northern hemisphere as the hoppers of the strand line, though a few species such as *Orchestia bottae* in Britain, have established themselves along the banks of rivers some distance from the sea. The terrestrial Amphipoda, however, are more closely allied to the strand-line species morphologically; so it appears that the Amphipoda have also colonized land directly rather than by fresh water.[269] Unfortunately little is known of the way in which water balance is maintained in these forms.

ISOPODA

Woodlice are largely cryptozoic forms confined to microhabitats of high humidity, though they differ widely in the extent to which they will withstand conditions exposing them to evaporative water loss. Their blood concentration is comparable with that of crabs but rather high by comparison with that of other terrestrial forms (table 11). Hence, gross water loss from the body must necessitate a larger absolute osmotic adjustment at the cellular level (see pp. 26 ff.) than is necessary in forms with lower blood concentration. Tolerance of loss of a large proportion of the body water is not therefore to be expected, though in fact *Ligia* has been found to tolerate a water loss sufficient to raise the blood concentration as high as $\triangle$ 3·48°C a rise of about 50% above the mean blood concentration of normal individuals.[402]

TABLE 11 (from Parry[402])

Genera	Mean $\triangle$,°C	Range, °C
Oniscus	1·04	0·81−1·25
Armadillidium	1·18	1·16−1·19
Porcellio	1·30	1·25−1·37

The water content of the body is maintained by a reduction of the permeability of the surface and by behavioural mechanisms designed to restrict the individuals to areas of high humidity. In the following series the animals show increasing adaptation to surviving dry conditions, though in moist situations several of the species may normally be found together. The permeability of the cuticles to water declines in the same order[165, 166]: *Ligia oceanica* > *Philosia muscorum* > *Cylisticus convexus* > *Oniscus asellus* > *Porcellio scaber* > *Armadillidium vulgari* > *A. nasutum*. The effectiveness of reductions in cuticular permeability is indicated by the fact that the rate of water loss per unit area in *Ligia* is about six times that of the least permeable form.[164] The rate of water loss varies with temperature, and there is a reasonable correspondence between saturation deficit of the air and the rate of water loss over a wide temperature range.

Reduction of water loss seems to be greater on the dorsal than the ventral side as more water is lost from the latter.[165] Decrease in permeability does not include the respiratory region and consequently, the greater the restriction of the surface permeability, the greater is the proportion of the total water loss which comes from the pleopod region.[165] In *Ligia* some 20% of the evaporative loss can be attributed to the pleopod region, but in *Porcellio* the proportion is 40%.[165] It is possible that surface losses of water might have been further restricted in isopods were it not for the fact that the cooling effect of transpiration serves as a temporary measure of protection against increases in external temperature proving fatal.   For example, *Ligia oceanica* is killed by exposure to saturated air at over 29°C for twenty-four hours, but it tolerates much higher temperatures for shorter periods if it is able to evaporate water.[165]   The body temperature depression may be quite considerable.   *Ligia* in a stream of dry air at 30°C may have a body temperature of only 24°C.[164]   Cooling the body by evaporation can obviously be only a short term measure, as in the long term the animal succumbs to water loss if not to heat-stroke; but nevertheless it may be of considerable survival value in giving the animal more time to seek a more favourable habitat. Reduction in surface permeability decreases the capacity to cool the body in this way, and coupled with this fact the species found to have the lowest permeability have the highest lethal temperatures.   Thus the upper lethal temperatures for twenty-four hours' exposure to water-saturated air follow a similar series: *Ligia oceanica* (29°C) < *Philosia muscorum* (30·5°C) < *Oniscus asellus* (31·5°C) < *Cylisticus convexus* (35°C) < *Porcellio scaber* (36°C) < *Armadillidium vulgari* and *A. nasutum* (37°C).[165]   Water is lost from the body in the faeces and urine also.   Faeces are moist when the animals have plentiful supply of water available but drier when the water is short.   It is presumed that glands present in the rectum are responsible for water re-absorption in the latter case as they are also in insects.   The excretory organs are maxillary glands, and no study of the urine volume or composition has yet been attempted.

Water lost by evaporation or in the urine or faeces is replaced normally by eating moist food.   If this is not available, woodlice resort to oral or anal drinking as they are unable to absorb water from the air (except in so far as, after previous desiccation, *Armadillidium nasutum* and *A. vulgari* can take up water from air with a relative humidity in excess of 98%.[165])   Water absorption through the integument is either absent or of little importance; for animals with the mouth and anus blocked show negligible weight changes when exposed to water.[490]   If a free water surface is available, water is taken into the gut both through the mouth and anus by *Ligia, Oniscus, Porcellio* and *Armadillidium*.   When water is available only as a moist plaster surface, desiccated *Oniscus, Porcellio* and *Armadillidium* can still take up water but *Ligia* is unable to gain weight under these conditions.[490]

3 : Moulting

The moulting cycle of the arthropods is one of the most fascinating aspects of their physiology. The presence of a stout cuticle has the obvious advantage that it provides protection for the body and an exoskeleton for muscle attachment; but it places severe restrictions on growth. Hence increase in body size has to occur in a series of steps associated with the casting of the old exoskeleton. At one time it was thought that moult (ecdysis) was a periodic event interrupting the normal life of the animal. Now it is recognized that the various stages of the moulting cycle are more or less continuous, the recovery from one moult being followed by storage of metabolic reserves and preparations for the next moult.

Immediately before and after the casting of the old exoskeleton, water is taken into the animal to expand the soft new cuticle. The consequent increase in size is sometimes loosely referred to as 'growth' but strictly the process is only one of expansion. True growth, the incorporation of new tissue, occurs during the later stages of the cycle.

Moult, though constituting only a short proportion of the whole cycle, is a period of some danger, and mortality is often high at this time.[273] The sources of hazard are threefold — mechanical, physiological and biological.

Mechanical difficulty may be experienced in withdrawing from the old cuticle, the expanded distal portions of the chelae of many decapods posing a particular problem.

Physiological problems arise from the considerable variations in the ionic ratios and total ion concentrations of the body fluids at moult, from the dilution produced by uptake of water into the cells, and from changes in the permeability of the body surface.

Finally, even if all these difficulties are overcome, the animal must still avoid the attentions of potential predators until the new cuticle is sufficiently hardened to make escape or defence possible. Anyone who has attempted to keep shore crabs in an aquarium will be aware of the fate suffered by any individual unfortunate enough to moult in the presence of its fellows.

A cyclical process as drastic as moult naturally involves the complicity of many other bodily processes. Indeed, as already mentioned in the introduction, structural or physiological modifications associated with moult have been found in such widely diverse systems as the central and peripheral nervous systems, ionic regulation, excretory system, cellular and general metabolism, blood system and reproductive cycle. Before discussing the role of these factors, however, the structure of the cuticle and stages in the moult cycle will first be considered.

1. Structure of the Cuticle

No single description will cover all the types of exoskeletal structure found in the Crustacea, since they range from the thin uncalcified flexible cuticles of the Branchiopoda to the thick, rigid carapace of decapods. In general, much of the organic material of the cuticle is the nitrogenous polysaccharide chitin; but, as this substance is relatively soft and flexible, the cuticle is strengthened by the deposition of calcium salts and/or by tanning (binding together of proteins by the formation of cross links between neighbouring molecules).

In the Decapoda the integument is subdivided into four main regions: epicuticle, pigmented layer, calcified layer and membranous layer (fig. 15C).

EPICUTICLE

The epicuticle is a more or less homogeneous layer of protein and lipid material, the protein being tanned by quinone cross links.[137, 510] Although lipid material is present, there is no evidence to suggest that a monolayer of aligned molecules, similar to that which forms the principal water barrier in the insect cuticle, is ever formed. Nevertheless, the epicuticle is largely responsible for the restricted permeability of the cuticle, at least in the uncalcified foregut membrane in *Homarus*.[566]   No chitin is found in the epicuticle, but this layer is sometimes calcified.

PIGMENTED LAYER

Below the epicuticle is a calcified chitin layer which, however, also contains tanned protein in its outer region.[137]   Viewed from

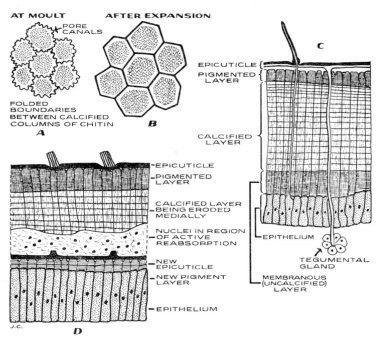

FIG. 15.   The structure of the cuticle.   *A. Cancer pagurus* cuticle in surface view, showing the appearance of the cuticular prisms before the expansion at moult.   *B*. Appearance of the prisms after moult expansion has occurred.   *C*. Vertical section of a decapod cuticle during intermoult.   *D*. The premoult condition, showing re-absorption of the old cuticle and the formation of the underlying new cuticle.   (*A*, *B* and *C* based on drawings by Dennell,[137] *D* based on a Figure by Yonge.[565])

the surface through the epicuticle, this layer gives the appearance of a series of hexagons (fig. 15*B*) the middle regions of which contain columns of chitin impregnated with calcium salts whilst tanned proteins are concentrated in the intercolumnar regions. The boundaries of the columns apparently mark the edges of the hypodermal cells originally responsible for the secretion of the layer. Pore canals, which originally contained protoplasmic filaments, extend vertically through the columns but are absent from the intercolumnar regions.[137, 144] Pigment granules are present in this layer as the name implies.

CALCAREOUS LAYER

An untanned chitinous layer more or less heavily impregnated with calcium salts lies below the pigmented layer. It forms much of the thickness of the exoskeleton.

MEMBRANOUS LAYER

This is an uncalcified, untanned chitinous zone lying immediately above the hypodermal cells.

The relative development of these various layers is by no means constant even within an individual. In soft regions, such as the intersegmental membranes, the cuticle is thinner than elsewhere, the degree of calcification is small and there is less evidence of tanning.[137] In contrast, where the exoskeleton is very strong not only is the calcification more extensive but there is also more tanning,[137] so that, paradoxically, regions often referred to as being 'heavily chitinized' in fact contain relatively less chitin than other, weaker, parts of the exoskeleton.[566]

A considerable degree of tanning is found in the cuticles of those forms in which a strong but light cuticle is necessary. Thus the jumping amphipod *Talitrus saltator*, a familiar inhabitant of the littoral strand line, shows considerable tanning of the epicuticle and pigmented layer but has little underlying calcification. In *Ligia oceanica* on the other hand lightness is less at a premium than in the hopping form and calcification is heavier and tanning less evident.

Localized areas of heavy tanning are also common where abrasive wear is likely to be considerable. The gnathobase setae of *Apus* and the tips of the walking legs of *Homarus* form diverse examples.[137]

2. Moulting Stages

Prior to ecdysis a new cuticle is partially formed below the existing exoskeleton and the medial layers of the latter are then largely re-absorbed (fig. 15D). The part that remains is cast off at moult and a rapid increase in body size occurs before the new cuticle is hardened. The various stages involved in the moulting cycle of crabs were first fully established by Drach[144] who studied *Cancer pagurus,* and with slight modifications these seem to be generally acceptable for all the Brachyura.[303] The characterization of the various stages is as follows:

Stage A   Immediately post-moult.   No feeding.

A_1 Exoskeleton so soft that the animal is unable to support itself on its limbs.   Weight increasing as water is still being absorbed.

A_2 Mineralization of the cuticle has begun and the animal can now stand but the exoskeleton is still soft.   Weight stable.   Water content of the whole animal is 86%.

Stage B   The main period of calcification of the new cuticle.   No feeding.

B_1 Secretion of the calcareous layer begins.   The meropodite and propodite of the legs can be bent without breaking.   Water content 85%.

B_2 Secretion of the exoskeleton continues.   Meropodite and propodite now crack if bent.   Water content 83%.

Stage C   The cuticle is hard, though calcification still continues during the early sub-stages.   Feeding is resumed.

C_1 The major period of tissue growth.   The opposite faces of the limbs give slightly if squeezed.   Water content 80%.

C_2 Tissue growth continues.   The limb cuticles give under slight pressure but crack if pressed firmly.   Water content 76%.

C_3 Integument rigid but calcification is still incomplete in the lateral and frontal part of the carapace. Water content 61%.

C_4 ' Intermoult stage.'   Calcification is complete and the membranous layer is laid down beneath the calcified zone. Metabolic reserves accumulating. Tissue growth complete. Water content 61%.

Stage D   Preparatory stage for subsequent moult.   Calcium re-absorption occurs and the outer layers of a new cuticle are secreted.   Feeding ceases and metabolic reserves are mobilized.   Activity declines after substage D_2.   Water content 59–61%.

D_1 The first sign of the impending moult is the appearance of new setae at the base of the old ones.   New epicuticle is secreted by the hypodermis.

D_2 The new pigmented layer is secreted.

D_3 Extensive re-absorption of calcium from the old cuticle occurs and this may result in cracking in places.

D_4 Re-absorption of calcium along the lines of dehiscence is complete, resulting in a gaping of the cuticle which makes possible the escape of the animal.   Water uptake begins.

Stage E   The animal escapes from the old exoskeleton and takes up water rapidly.

The portion of the cycle accounted for by each stage differs according to the species involved but values for *Panulirus* may be taken as an approximate guide.   These are: A 2%, B 8%, C 71% and D 19%.[509]

The length of the cycle varies with the age of the animal, species and time of year and may also be influenced by the temperature and nutritional state as the following examples illustrate. Low temperatures often inhibit moulting; thus moulting does not occur in the crab *Pachygrapsus* when the temperature is below 8.5°C.[452]   Bermudan individuals of *Panulirus argus* are even more sensitive to low temperatures, moulting being virtually inhibited below 17°C.[509]   Conversely, the moulting of the land crab *Gecarcinus* is delayed if it is exposed to abnormally high temperatures.   The effect of both low and high temperature inhibition is to lengthen the intermoult stages.

Young animals tend to moult more frequently than older ones. Thus the average summer intermoult period of *Panulirus* with a carapace length of 80 mm is about sixty-five days whereas that of smaller ones with a carapace length of 20 mm is only about forty days.   As a result of the shorter length of the cycle the young *Panulirus* would be expected to complete five moults a year, but the larger ones only three or four.

In some anomuran crabs exposure to high temperatures during larval development has the remarkable effect of increasing the number of moult stages prior to metamorphosis.[47]   In Brachyura on the other hand the number of premetamorphic moults is immutable.[47]

Starvation inhibits the onset of moulting in *Pachygrapsus*,[452] but, curiously, has the opposite effect on *Crangon* and *Leander*.[452]

The rate of moulting shows marked specific differences.   Thus by contrast to the cycle of *Panulirus* which lasts some sixty days moult occurs every fourteen to eighteen days in the prawn *Leander xiphias*[461] and every two to three days in *Balanus*.[112, 114]

Sexual differences in the cycle length have been noted in *Leander*, the C_4 stage occupying a smaller portion of the total cycle and the D stage having a shorter absolute length in the male than in the female.   In consequence, the male moults somewhat more frequently than the female.[461]

When the whole cycle is of short duration, the individual phases also are, naturally, compressed.   Processes which normally take days or hours in crabs may be completed very rapidly in smaller forms.   To take an extreme example, water absorption is not complete until some six to twelve hours after moult in *Maia squinado*, *Carcinus maenas* and *Cancer pagurus*: but in *Daphnia* the size increase is complete within one minute of moult.[163]

Size increase in young animals may be considerable but with increasing age there tends to be a decrease in the proportionate expansion at each moult.   When *Daphnia* are kept in unfavourable conditions, there can even be a decrease in size at moult.[223]

In the decapods differentiation has been made between two types of moulting cycle, anecdysis and diecdysis.[303] Anecdysis occurs in forms such as *Cambarus*, *Uca*, *Maia* and *Panulirus*, which moult on a seasonal or annual basis.   In the cycle of these forms there is a long C_4 stage as outlined above. By contrast, in diecdysis, there is no prolonged intermoult stage, one moulting cycle merging into the next in such a manner that the C stage is generally somewhat shorter than the subsequent D stage. Diecdysis is characteristic of forms such as *Leander*, *Ligia* and *Lysmata*, which moult throughout the year; but some crabs, including *Pachygrapsus* and *Carcinus* tend to have a number of diecdysic moults during the summer followed by an anecdysic cycle, with a longer overwintering intermoult.[303] An extreme example of anecdysis is seen in *Maia*, which moults only in July and August except when very small.   A number of crabs do ultimately reach their maximum size and then cease moulting. This so-called ' terminal anecdysis ' has been observed in *Maia*, *Carcinus* and *Callinectes*.   It does not occur in prawns or *Homarus*.   Feeding in anecdysic forms occurs only in the C and early D stages.   This situation would clearly not be practicable in diecdysic forms because of the curtailment of the C stage, and in fact the diecdysic *Leander* has been observed to continue feeding until the equivalent of the D_4 stage immediately preceding moult and to resume feeding during the B stage.[461]   Dependence on food reserves is thus less in diecdysic forms since the non-feeding period is shorter; and appropriately the diecdysic *Leander* stores less glycogen in the hepatopancreas than do the anecdysic brachyurans.[461]

3. Development of the New Cuticle

Before any formation of a new cuticle can occur the old cuticle must be detached from the hypodermis.   In *Panulirus* this occurs some ten to fourteen days before moult.[510]   The hypodermal cells then increase in number, elongate, and secrete the epicuticle and pigmented layers of the new cuticle (fig. 15*D*).[510, 565] In *Carcinus* and *Panulirus* the epicuticle is tanned by quinones immediately after its formation.[307, 510]   The tanned lipoprotein layer therefore separates the chitinous layers of the old cuticle from the chitin containing new pigmented layer.   This separation

6

is probably of considerable importance in protecting the newly formed chitin during the re-absorption of material from the old cuticle which now occurs.[565]   Fluid appears between the new and the old cuticles, and in this cell-nuclei can be seen that probably belong to amoeboid cells involved in breaking down the old cuticle.[565]   Digestion of the old cuticle continues until all the membranous layer and much of the calcareous layer have been destroyed.   The space formed allows for the easy separation of the old and the new cuticles necessary if the animal is to free itself of the former.   Re-absorption is most complete along the ecdysial lines, but is also considerable at the base of those limbs where the distal region is expanded.   The latter is an obvious requirement if the tissues of the distal region are to be withdrawn through the narrow neck at the base of the appendage.

At the time of moult the new epicuticle in *Panulirus* is complete, and the pigmented layer has been secreted but not tanned so that it retains a considerable degree of flexibility.   The body water content starts to rise shortly before the escape from the old cuticle.   The animal now undergoes a variety of flexing and lunging movements as it endeavours to free itself.   Finally, after selecting a secluded spot, it flexes the abdomen and emerges through the intersegmental membrane separating the thorax and abdomen.[509]

The body is rapidly expanded during and following moult. One feature which is of particular importance in enabling the lifting of the thoracic cuticle prior to emergence is a swelling of the stomach.   This is brought about by the swallowing of the medium in *Cancer pagurus* and *Maia squinado*[144] but in *Panulirus* no fluid is present in the gut.   Expansion of the stomach of this last animal is brought about instead by gas which appears between the old stomach lining and the new one.   Gastroliths, lens shaped concretions of calcium carbonate, develop between the old and new stomadael linings as calcium is re-absorbed from the cuticle in the moulting Reptantia and semi-terrestrial brachyuran, *Sesarma*[387]; but it is still not certain whether they are involved in the evolution of the gas or not.[565]

Most Crustacea moult the whole exoskeleton at once, but in isopods the fore and hind parts of the old cuticle are cast off separately, often several hours or even days apart, so that the animals may retain a greater degree of motility at this critical time than is commonly the case.

Moult in many species occurs during the hours of darkness, thus providing a measure of protection against predators whilst the animal is helpless. If exposed to continuous light such forms may delay moulting for a time.[42]

A virtue is apparently made of the difficulty of escaping from the old cuticle by the cladoceran, *Alonopsis*. This animal, which moults incompletely and so carries a number of its previous exoskeletons about with it, is found to be much less common in the gut of the predatory worm *Chaetogaster* than might be expected on the basis of its abundance in Lake Windermere. This observation has led to the interesting suggestion that it may be able to slip out and escape when the old cuticle is grasped by the worm,[224] after the manner of a stage felon leaving his jacket in the hands of a comic policeman.

TABLE 12 (from Robertson[456])

	Water uptake as % premoult wt.	Water % intermoult wt.	Water % post moult wt.	Cast cuticle % premoult wt.
Maia squinado[144]	125·0	69·2	90·1	23·3
Carcinus maenas[456]	66·3	65·4	86·9	23·9
Uca pugilator	42·9	67·9	84·3	50·9
Cancer pagurus[144]	94·0	62·2	86·1	26·6
Hemigrapsus oregonensis[388]	44·9	67·7	84·9	
Pachygrapsus crassipes[388]	33·9	66·8	83·8	18·0
Panulirus argus[509]	18·8	71·0	82·4	20·1

Water uptake by the blood and tissues starts just before emergence. In *Panulirus* most of this uptake is apparently across the body surface,[509] but in *Maia squinado*,[144] *Cancer pagurus*[144] and *Carcinus maenas*[456] it is largely from the gut. In view of this difference it may be significant that the total uptake by *Panulirus* is very much less than that in the brachyurans.[456] As water enters the body, some of it is taken up into the cells and the remainder serves to increase the blood volume. The rise in the water content of the hepatopancreas cells of *Carcinus* at moult suggests that about one third of the water taken into the body passes into the cells,[456] and on the basis of determinations of the drop in blood protein content it is estimated that the blood volume is increased to some 240% of the premoult level.[456]   Water uptake serves to stretch the new cuticle, and the irregular borders of the columns of calcareous material in the pigmented layer straighten to give a hexagonal appearance in surface view (fig. 15A).

One of the factors contributing to the uptake of water by *Carcinus* is a rise in the osmotic pressure of the blood during the premoult stages.   Intermoult animals in sea water have a blood concentration within 1% of that in the medium but during premoult the blood may be as much as 8% more concentrated than the medium.[456]   Water will therefore tend to be drawn into the body by osmosis.   The fast water uptake at moult decreases the osmotic gradient by diluting the blood, and within twenty-four hours of ecdysis the concentration is only about 3% above that of the medium.   Much of the premoult rise in concentration is due to increases in the concentration of ionic constituents, the percentage increases in calcium and magnesium being particularly striking.[387, 456]

The rise in blood calcium level is presumed to be due to the re-absorption of calcium from the cuticle[456] during the D stages, despite the fact that much calcium is also being deposited in the hepatopancreas at this time.   In post-moult animals a drop in blood calcium is contemporaneous with the calcification of the new cuticle.   There is, however, insufficient calcium in the blood initially to account for the degree of calcification of the new cuticle; so calcium must also be taken up from the medium.[456]

It is obviously desirable that immediately after moulting animals should display as little muscular activity as possible if the new cuticle is not to suffer distortion.   The high levels of calcium and magnesium ions in the blood at this time may counteract this risk by tending to depress nerve and muscle excitability.   It is probable, however, that the function of the special inhibitory nerve supply to the appendages (p. 162) is also to depress activity, though this has yet to be demonstrated experimentally.

As we have already seen, about one-third of the water taken in by *Carcinus* at the time of moult passes into the cells.   If the inorganic ion concentration of the cells is to be maintained at the normal level, this uptake of water must be accompanied by an increase in the osmotically active organic components of the cells. This is achieved by increasing the free amino-acid level.   In the case of *Carcinus* the major participants in this increase are glutamic acid and proline.   Glycine, which plays a major role in the adjustment of cell osmotic pressure when the blood concentration of inter-moult animals is changed (p. 27), here takes little part.[150] It has been suggested that the necessity of adjusting the cellular amino-acid level to take account of water uptake at moult may have formed the basis for mechanism adopted by euryhaline Crustacea to match their cell concentration to changes in blood concentration.   However, as the non-arthropod brackish-water species also utilize amino-acids for osmotic adjustment of their cells, this cannot be regarded as a vital pre-adaptional preliminary to the use of amino-acids for this purpose.

4. Control of the Moult Cycle

Moulting is controlled by hormones and before the mechanisms involved in initiating the process are discussed it is therefore necessary to digress to consider the nature of hormones and the organs from which they are released.

4 : Hormones

Hormones are blood-borne chemical agents which act on cells at a point distant from the site where they are produced. Aspects of the physiology regulated by hormones are very varied, including —in addition to the moult cycle—such diverse factors as heart rate, sugar metabolism, gonad maturation and the dispersion of pigment in chromatophores.

The tissues producing hormones in Crustacea fall readily into two categories : (1) specialized cells of the central nervous system (neuro-secretory cells), (2) tissues of non-nervous origin. In general the secretion of the latter group is stimulated or depressed by hormones from the former. Most of the humoral activity is thus ultimately under the control of the central nervous system. The land crab *Gecarcinus* provides an excellent illustration of this dependence on the central nervous system, since this animal can delay ecdysis for a time if it is in circumstances (bright light, high temperature, or in the company of its fellows) likely to prove deleterious if it were to moult.[43]

1. Neurosecretory Cells

All nerve axons release substances which excite or depress neighbouring tissues. To some extent, therefore, all nerves may be regarded as secretory. The term ' neurosecretory ' is, however, reserved for those cells which are specialized for the storage and release of humoral agents. As such agents are carried by the blood, the axons of neurosecretory cells always terminate in close proximity to blood sinuses, instead of on some effector cell as do ordinary nerve axons. Granules which stain densely with

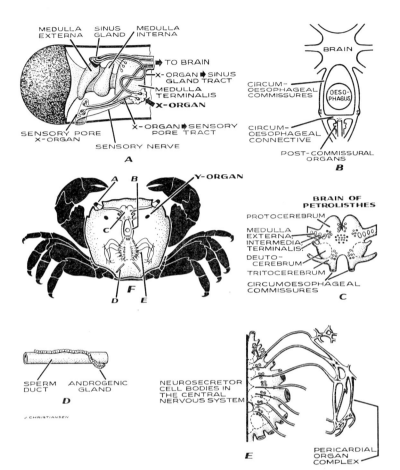

FIG. 16. Some important sites of hormone formation and release in Crustacea. *A.* A generalized Crustacean showing the approximate sites of various humoral organs. (*Note:* The precise form or position of the organs often differs from group to group.) *B.* A prawn eyestalk, showing the relation between sinus gland, X-organ and the sensory pore organ. (*Note:* The sensory pore organ is absent in Brachyura.) *C.* An anomuran (*Petrolisthes*) brain to illustrate the site of neurosecretory bodies. *D.* Post-commissural organ (from the prawn *Leander*). *E.* The androgenic gland (from *Callinectes*) lying along the sperm duct. *F.* The pericardial organ complex of a Brachyuran. *G.* The Y-organ. This structure is composed of a mass of more or less equal-sized small cells. (*B* and *D* from Carlisle and Knowles,[94] *C* from Kurup,[316] *E* from King [293] and *F* from Maynard.[369])

77

Gomori-chrome haematoxylin are often found in neurosecretory cells; and, as these granules originate near the nucleus and then move along the axon, they are presumed to be either the hormonal material itself or a precursor.

The cell bodies of neurosecretory cells occur in various parts of the central nervous system but the sites of hormone release are more localized. Aggregations of the axon terminations are recognized in several areas, the most marked being the pericardial organs, the post-commissural organs, the sinus glands and Hanstrom's sensory pore organ. The location of these structures in a generalized decapod is as in fig. 16.

Active extracts have been obtained from many other regions of the nervous system, but it is still doubtful if these represent natural hormone sources. If they do, it is probable that they are the precursors of the true hormones in transit in nerve axons. At present, only three tissues of non-nervous origin are known to produce hormones, though doubtless others remain to be discovered. The three tissues whose humoral role has been established are the Y organ, the androgenic gland, and the ovary.

The hormones produced by these glands can be regarded as primary effectors since they act directly on other body tissues. The Y organ initiates moult, the androgenic gland controls the primary and secondary sexual characters of the male, and the ovary determines secondary sexual characters in the female. By contrast, at least some of the hormones produced by the neurosecretory glands are secondary effectors, since the alteration in any physiological process due to their secretion is not the result of a direct action of the hormone on the tissue but is due to its control of the rate of secretion of one of the three glands mentioned above.

2. The Androgenic Gland

The androgenic gland is responsible for producing a hormone or hormones which determine all the primary and secondary sexual characters of the male.[101]

Originally discovered in the males of the amphipod *Orchestia*,[100] the gland has now been located in nearly all the super-orders of higher Crustacea.[101] Usually it consists of a strand of cells lying along the vas deferens near its termination[101] (fig. 16) and so between the coxopodite muscles of the last thoracic leg; but in some isopods it lies close to the testicular tubules.[333]   It is absent in females.

Implantation of the androgenic gland into a young female prevents the development of female secondary sexual characters, such as oostegites, and results in mascularization including a change in the appendages towards the male condition.   The effect of the hormone is clearly a direct one, since the ovary can be removed prior to implantation without altering the result.   If the ovary is not removed, it ceases to produce eggs and instead forms sperm which are fertile and can be used artificially to fertilize normal eggs.[101]   Females with implanted androgenic glands display normal male behaviour and will mate with normal females. They cannot, however, fertilize the females as, although a vas deferens with a lumen is developed, it is not functional.[101]   Injection of testes without androgenic gland into females has no mascularizing effect.   Reciprocal transplants of ovaries into males give further proof that control of male characteristics is governed solely by the androgenic gland.   Thus, when an ovary is transplanted into a normal male, it is converted into a testis. When an ovary is put into a male from which the androgenic gland has been removed, however, it remains as an ovary.[101]

Finally, removal of the androgenic gland from a male can result in a progressive loss of male secondary characters, and in at least one case, *Orchestia montagui*, the testis may become converted into an ovary about a month after the operation.[101]   This does not happen, however, in *O. gammarella*.[101]

In the isopod *Porcellio* the male secondary sexual characters controlled by the gland include the brushes of pereiopods 1–4, the gonopods, the genital appendix and the uropod exopodite.[346]

In some Crustacea the androgenic gland displays seasonal activity, being smaller in winter than in summer ; and there is an indication that its activity is controlled by an inhibitory hormone from the eyestalk.[101, 134]

Injections of steroids, which in mammals have androgenic effects, have no such action in Crustacea[101]; and, since the histological appearance of the gland resembles that of vertebrate protein-producing cells, it seems possible that the effective hormone is a protein or peptide.[101, 293]

3. The Ovary

The ovary is responsible for the production of hormones determining the temporary and permanent sexual characters of the female. The female of the amphipod, *Orchestia gammarella*, has oostegites as a permanent secondary sexual character and as a temporary feature long hairs develop on the margin of the oostegites at the moult preceding egg laying. If the ovary is removed then these hairs revert to the juvenile condition at the next moult. Furthermore, if an ovary is implanted into a castrated female then ovigerous hairs will once more develop when yolk deposition occurs in the implanted organ. Unlike the testis, therefore, the ovary can be regarded as a humoral organ in its own right.[101]

SEX DETERMINATION

In the experiment described above, where eggs of a normal female *Orchestia* were fertilized by sperm produced by a female mascularized by implantation of androgenic glands, it was found that the resultant offspring included both males and females. The most likely explanation of this is that the female is the heterogametic sex (XY) and the male the homogametic sex (XX).

The juvenile amphipod has an undifferentiated strand of tissue which will become the future gonad and in all individuals the anlagen of both the vas deferens and oviduct are present. The genetic composition of the individual determines whether or not the androgenic gland will be differentiated from this primordial genetic tissue. In the male it develops and its hormonal activity results in the formation of the remaining gonadial tissue into a testis. In the female the androgenic gland does not develop and the gonad becomes the ovary.[101]

Despite this relationship between the genetic composition and the sex, it appears that the sex ratio in the population as a whole can be made to vary by external conditions. Thus, when *Gammarus duebeni* parents are kept continuously at low temperatures (below 5°C), they are said to produce only male offspring; between 5 and 8°C the offspring are of both sexes; and above 8°C all are female.[294] Field observations support these findings, since there is a preponderance of males in the early spring months, and of females later in the summer.[294] The manner in which temperature affects the sex ratio is as yet unknown and presents a somewhat intriguing problem.

A number of decapod species are known to change sex during the life of the individual. One species, the Thalassid, *Calocaris macandreae*, is a functional hermaphrodite [67b] the others are all protandrous hermaphrodites, being male at first and female later. This latter group includes the following prawns, *Pandalus borealis*, *P. danae*, *P. montagui*, *P. kessleri*, *P. platyceros*, *P. hypsinotus*, *Lysmata seticaudata*, *L. nilita*, *Chlorismus antarcticus*, *Campylonotus rathbunae*, *C. semistriatus* and *Pandalopsis dispar*.[561c]

An ovary-inhibiting hormone is present in the sinus gland and pars distalis X-organ of *Palaemon* and *Lysmata*, and there is good evidence that a similar hormone delays the onset of the female condition in *Pandalus borealis*, one of the forms showing sex reversal.[92c]

4. The Y Organ (Ventral Gland)

The Y organs, first discovered by Gabe,[215] bear a close resemblance to the thoracic or prothoracic glands of insects.[216] In brachyuran crabs they are located under the ventral hypodermis in proximity to the mandibular external adductor muscle and in a comparable situation in many other forms.

The hormone produced by the gland is responsible for initiating the stages leading to ecdysis, and if the Y organs are removed from crabs, the animals fail to undergo any further moults. Proof of the hormonal nature of moult control is provided by the fact that implantation of several Y organs into crabs lacking these structures can restore their ability to moult.[162] During intermoult the cells of the glands contain material which stains with aldehyde fuchsin, but this is absent in freshly moulted animals.[159] Presumably this is the stored precursor of the hormone.

The secretory activity of the Y organs is not under nervous control but is determined by the presence or absence in the blood reaching the gland of a hormone (the moult inhibiting hormone) from the sinus gland in the eyestalks (p. 84). When the inhibiting action of this substance is removed, as by ablation of the eyestalks, the onset of moult is precipitated. The interaction of the two humoral systems is elegantly shown by experiments in which both the eyestalks and the Y organs are removed from *Carcinus* which have undergone their terminal ecdysis and which would not therefore be expected to moult again.   When the eyestalks (or the X organ plus sinus gland alone) are removed, about two-thirds of the crabs moult, in the case of *Carcinus*,[92] and even greater proportions in *Uca, Sesarma, Eriocheir,* and *Gecarcinus*.[408] Removal of both eyestalks and Y organs, however, does not result in moulting.[408]

Moult can go to completion, however, if the Y organs are removed after the early D stages (D_0 and D_1), thus indicating that their role in the process is associated with the initiation of proecdysis and not primarily with the later stages.

Regenerating limbs have proved a valuable material for the investigation of the action of the moulting hormone at the tissue level.   When a limb is autotomized a limb bud subsequently develops.   This ' basal growth stage ' is, however, restricted to the initial reorganization and development of the new tissues and when this is complete growth stops.   The limb bud shows no further size increase after this stage until either the animal's eyestalks are removed to stimulate Y organ hormone release or alternatively it naturally proceeds into proecdysis.[39]   The limb bud then starts to increase in size.   Additional support that this resurgence of growth is brought about by Y organ hormone seems to be given by the fact that it can be checked by implanting eyestalk extract or by exposing the crabs to a bright light[43] or the company of other crabs (fig. 17).   Such reflex cessation of further preparation of moult, even after proecdysis has begun, is presumed to be brought about by central nervous stimulation of further release of the moult-inhibiting hormone.[39, 449]   The ecological advantage of this physiological adaptation is, clearly, that the moult is most likely to take place only under optimum conditions, i.e. in darkness, in solitude, and in a moist burrow.

The rapid growth of an already formed tissue is associated with cell division and it is intriguing therefore that it has been found that mitosis increases markedly during those stages of proecdysis which are sensitive to the presence of the hormone,[501] though it cannot of course be stated that the hormone directly stimulates cell division on this evidence alone.   Nevertheless, since the Y organ hormone has been found to be necessary for the initial stages of gonad maturation as well as limb bud development,[16, 133] it at least appears to be " required for further development of already organised tissue involving cellular duplication ".[408]

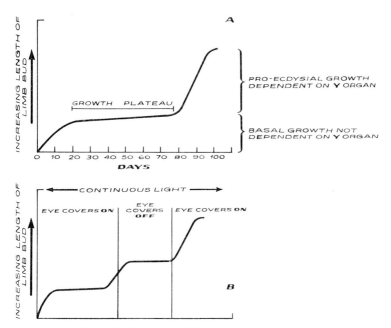

FIG. 17. Stages in the regenerative growth of a limb bud of *Gecarcinus lateralis* following removal of a thoracic leg: *A*. The normal stages, including initial basal growth, a plateau during the remainder of the intermoult stage, and then pro-ecdysial growth during the D_0 and D_1 stages. *B*. The inhibition of pro-ecdysial growth by exposure to bright light. A similar plateau occurs after the onset of pro-ecdysial growth if another crab is introduced into the cage. (Modified from Bliss and Boyer.[43])

Another feature observed after injection of Y organ extracts into crabs is a sharp rise in the blood calcium level,[92] but this is likely to be a secondary rather than a primary response to the hormone.

Histologically the Y organ is very similar to the prothoracic gland of insects which fulfils the same purpose, initiating moulting by the hormone ecdysone. Insect ecdysone will indeed also cause

Crustacea to moult,[286] and conversely crustacean extracts are effective on insect larvae.[286b]   Very little hormone is stored in the Y organ of intermoult animals; more being present in the blood than in the gland at this time.[92b]   Homology in structure and function between the Y organ and thoracic gland of insects is to be expected, since both are derived from ectodermal ventral glands.[542]

5. The Sinus Gland and X Organ

In most stalk-eyed Crustacea the sinus gland is situated about two-thirds of the way along the eyestalk (fig. 16) and lies on the dorsal surface of the optic ganglion.   A comparable structure lies within the head in the Anomura[316] and sessile-eyed Malacostraca.[12] Despite its name it is not, strictly speaking, a gland since it is not composed of secretory cells.   Rather it is a storage release site for hormonal material formed elsewhere in the central nervous system and then transmitted at the rate of a few millimetres per day along the nerve axons to the sinus gland.   The cell bodies of the axons whose swollen ends form the sinus gland are located in various regions, including the X organ, the eyestalk, and the more central regions of the brain (proto-, deutero- and trito-cerebrum).[92]

On the appropriate stimulus, probably impulses in the axon,[39] hormone is released into the blood sinuses which the axon terminations surround.

The sinus gland, supplied as it is by axons from several sources, may be the site of the release of more than one hormone. The removal of the eyestalks is a simple operation and consequently the part played by the gland in regulating various metabolic processes has been considerably studied.   The following functions are claimed[170] for hormones released by the tissue: (1) moult inhibition, (2) control of sugar metabolism, (3) control of metabolic rate, (4) depression of gonad development, (5) control of pigment dispersion, (6) control of body protein metabolism, (7) control of water metabolism, and (8) control of heart rate.

Histologically different regions can be made out in the sinus gland but as only six have so far been located,[418] any simple correlation with the above list of functions is therefore ruled out unless possibly one hormone can have several physiological effects.

MOULT INHIBITION

The inhibitory control of the sinus gland over the activity of the Y organ was mentioned in the discussion of the latter (p. 81). The hormone responsible apparently comes from the X organ initially, since removal of the latter alone results in Y organ activity and abnormally accelerates moult, though not, as might be expected, quite so rapidly as removal of both X organ and the sinus gland.[406] Re-implantation of the sinus gland X organ complex into crabs lacking eyestalks delays the onset of the precocious moult.[406]

Two types of cell are present in the X organ and those designated 'type 2' stop secreting four or five days before normal moulting in crabs and restart a few days after ecdysis.[158] This is the type of behaviour to be expected of the cells responsible for producing the hormone that inhibits Y organ activity at times other than moult, and it is probable that they represent the source of the hormone.

In larval decapods the sinus gland is formed quite late in development (the fifth larval stage in *Palaemonetes*), and it does not accumulate stainable storage material at all during larval life. It is presumably no coincidence, therefore, that the larvae pass rapidly from one moult to the next with no prolonged intermoult, and that removal of the eyestalk has no effect on the rate of moulting.[266] It has been suggested[90] that species showing diecdysis may also have inactive X organs allowing them to pass rapidly from one moult to the next, in contrast with anecdysic forms and although there is no direct support for this[408] in decapods, it is possible that there is no inhibitor in the barnacle *Pollicipes polymerus*.[22]

It was mentioned earlier that certain crabs ultimately cease to moult and thus pass into a state known as terminal anecdysis. This situation may arise in one of two ways both in theory and practice. Either, as in *Carcinus*, there is a continuous release of inhibitor by the sinus gland so that the Y organ can never secrete the moulting hormone, or alternatively, as in *Maia*, the Y organ itself may regress. *Carcinus* in terminal anecdysis can be caused to moult again by removing the eyestalks but this operation obviously does not cause *Maia* to moult.[92]

CONTROL OF SUGAR METABOLISM

Exposure of some Crustacea to stress, e.g. by injecting distilled water into the haemocoel, causes an increase in the blood sugar level. This hyperglycaemia does not occur in crayfish, however, if the sinus glands have first been removed,[301] suggesting that the latter may release a controlling factor. In confirmation of this, it is found that injections of sea-water extracts of the eyestalk into *Uca*, *Callinectes*[1] and *Panulirus*[462] cause an increase in blood glucose. Most of the active principal in the eyestalk appears to be located in the sinus gland.[1] Removal of the whole eyestalk might therefore be expected to result in lowered blood sugar levels. This does indeed happen in *Panulirus*,[462] but not in *Libinia*,[300] *Astacus*[301] or *Callinectes*.[1]

CONTROL OF METABOLIC RATE

Bilateral removal of the sinus glands of the crayfish *Procambarus* results in an increase in the oxygen consumption of the animal, though there is little change if only one gland is removed.[473] Since bilateral removal of the Y organs has no effect on oxygen consumption, it seems likely that the effect of the removal of the eyestalks can be regarded as acting directly and not via the Y organ activity.[384] The converse experiment of injecting eyestalk extract into eyestalkless crabs (*Uca*) has the predicted effect of depressing the raised rate of oxygen consumption.[170]

It seems likely that the cell bodies of the axons releasing the hormone responsible for depressing oxygen consumption do not lie in the eyestalk itself, as the effect of eyestalk removal tends to vary according to whether the wound is cauterized or not.[504] Eyestalk removal from *Uca* results in the usual increase in oxygen consumption in uncauterized preparations but when the wound is cauterized there is no increase and there may be even a slight fall.

When *Orconectes virilis* is exposed to a bright light the eyestalks release a hormone which depresses the activity of the animal.[451] Conceivably this could be the same hormone as that responsible for controlling the oxygen consumption. It is also likely that a similar hormone plays a major part in adjustments to metabolism which occur in temperature adaptation (see pp. 136 ff).

In addition to increasing the oxygen consumption, removal of the eyestalk also depresses the respiratory quotient.

THE DEPRESSION OF GONAD DEVELOPMENT

The inhibitory control of the neurosecretory cells of the eye-stalk over the gonads is indicated by the rapid maturation of the latter that occurs when the eyestalks are removed. Thus de-stalking juvenile females results in the development of the ovaries.[524] This effect is particularly well shown in *Palaemon*, (*Leander*) *serratus*, where the ovaries of de-stalked animals rapidly increase to some thirteen times the size of those of normal immature controls.[394] Maturation of the eggs and egg-laying can

FIG. 18. The effect of removing the eyestalks of juvenile *Uca* on the maturation of the gonads. *A*. Gonad size one month after removal of the eyestalks. *B*. Gonad size in normal controls. (Redrawn from Brown and Jones.[55])

then follow in animals which in age are still immature.[394, 395, 396] Similar effects occur in *Cambarus immunis*,[54] and *Uca pugilator*[55] after removal of the eyestalks (fig. 18).

The ovary-inhibiting hormone apparently originates from the medulla terminalis X organ.[89] Implantation of the sinus gland into eyestalkless juveniles of *Uca* has some, but not a complete, effect of decreasing the precocious development of the ovaries. Removal of the eyestalks of immature males of *Carcinus* results in hypertrophy of the androgenic gland (*q.v.*) and increased development of the testes.[131, 132, 134, 135]

Mention has already been made of the fact that the Y organ hormone plays a part in the maturation of the gonads by stimulating cell division. Since the gonad-inhibiting hormone also arises

7

in the X organ,[90] as does the moult-inhibiting hormone, the possibility must be considered that they are one and the same and that gonad inhibition is achieved by inhibition of Y organ activity. It is generally considered, however, that they are in fact separate hormones,[102] since the gonad-inhibiting hormone seems to be associated principally with the prevention of yolk deposition rather than of cell multiplication.

6. The Control of Chromatophores

In the Malacostraca pigments of various types are present in specialized cells, the chromatophores. The degree of expansion of the pigments within the chromatophores is controlled by hormones released from the eyestalk and also from other parts of the body.

The chromatophores may be divided into functionally distinct groups:

(1) The pigment cells of the eye.
(2) The hypodermal pigment cells.

The pigment cells of the eye are involved in the adaptation to different light intensities. The hypodermal cells are responsible for matching the animal to its background i.e. camouflaging it, or alternatively for thermo-regulation.

THE EYE PIGMENTS

The pigments of the eye are located in three groups of cells which are termed the *distal pigment layer*, *proximal layer* and *reflecting layer* (fig. 19).

The distal pigment cells (two) form a sheath round the crystalline cone and tract, the proximal pigment is located in the retinular cells (eight) themselves, and the reflecting pigment in cells—fewer than the ommatidia—around the base of the retinular cells. In prawns all three pigments migrate during the adaptation of the eye to differing light intensities. In light-adapted eyes the distal pigment moves medially and is dispersed. The proximal pigment is dispersed in the retinuli above the level of the basement membrane, and the reflecting pigment remains below the basement membrane. By contrast, in the dark-adapted eye the distal

pigment is concentrated and moved distally, the proximal pigment
is withdrawn below the level of the basement membrane, and the
reflecting pigment moves above the basement membrane so that

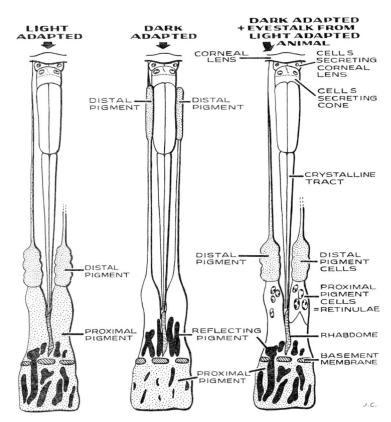

FIG. 19. The effect of hormones on pigment migration in the
chromatophores of the eye. (1) Light-adapted eye. (2) Dark-
adapted eye. (3) Eye of dark-adapted animal injected with extract
from a light-adapted animal. Note that the extract causes movement of distal and reflecting pigments but not of the proximal
pigment. (From Kleinholtz.[297])

its reflecting function is increased. The distal and proximal
pigments are black, the reflecting pigment white. Modifications
of this scheme are found in other species. Thus the reflecting

pigment of *Cambarus*[545] and *Eupagurus*[125] does not migrate. The functions of the pigment in adaptation to light and dark are discussed on pages 237 ff.

The migration of the proximal pigment in response to a change in light intensity is considerably more rapid than that of the distal pigment,[125] and there is generally no change in the pigment following injection of eyestalk extracts.[297] It is presumed, therefore, that the dispersion of this pigment is normally brought about directly by a photo-mechanical effect and not by the intervention of a hormone. Some recent evidence, however, does suggest that the effect may be modified by a hormone, at least during dark-adaptation.[184]

The distal pigment is under the influence of hormones released from the sinus gland and the central nervous system. When crude extracts of the eyestalk[297] and also of the ventral nerve cord circum-oesophageal connectives and super-oesophageal ganglion are injected into dark-adapted animals,[63] they bring about the light-adapted condition fig. 19. However, by treatment of the eyestalk extract by electrophoresis at pH 9, it is possible to separate another hormone which, when injected into light-adapted animals, produces the dark-adapted state of the distal pigments.[184] The light-adapting and dark-adapting hormones are both present in extracts of the sinus gland but since the total activity of extracts of the optic ganglion plus sinus gland is similar to that of the sinus gland alone it may be presumed that the source of the hormones released by the sinus gland is not in the eyestalk, i.e. is not the ganglionic X organ. The trito-cerebral commissures which link the two circum-oesophageal connectives behind the oesophagus are a potent source of the dark-adapting hormone. By contrast brains plus circum-oesophageal connectives but without the trito-cerebral commissures contain only light-adapting hormone.[186] It is probable therefore that the sinus gland light- and dark-adapting hormones have their origin in the brain. The position of the distal pigment is determined by the balance between the light- and dark-adapting hormones since the effects are mutually antagonistic.[184]

Synthesis of both hormones apparently continues when the animals are kept for an extended period in darkness since the amount of light-adapting hormone the supra-oesophageal ganglion then increases.[185]   Both the light-adapting and dark-adapting hormones are inactivated by trypsin suggesting that they are polypeptides.[185]

THE HYPODERMAL CHROMATOPHORES

The chromatophores of the hypodermis and sub-hypodermis function in matching an animal to its background and in thermo-regulation.   They are much branched cells or syncytia ramifying between neighbouring cells.   One, two, three, or four pigments may be present in each though usually spatially separated in different branches of the syncytium.   Individual pigments can be dispersed or concentrated independently of others in the same chromatophore.   The most common pigments are black, dark brown, red and white, though yellow and blue may also be present in some forms.   Brown and black pigments of the Brachyura are melanins (polymerized indole quinones produced by the action of tyrosinase on amino acids, particularly tyrosine.)   Red pigments are astaxanthin or other carotinoids.

Two factors are involved in adaptation of the coloration to match the animal's surroundings:

(1) Morphological colour change.
(2) Physiological colour change.

Morphological Colour Adaptation

This involves a change in the amount of pigment present and degree of ramification of the chromatophores.   Thus in *Palaemonetes vulgaris* the quantities of red, blue and yellow pigments decrease when the animal is kept continuously on a white background and the red and blue increase when it is on a black background.   The process is a slow one taking many days before completion.[53]   In crabs morphological colour change is brought about by increasing the number of chromatophores per unit area, by increasing the amount of pigment per chromatophore, or by a combination of these methods.[224b]

Physiological Colour Variation

Physiological colour variation is a rapid process not involving change in shape of the cell or in the total amount of pigment present.   Instead the degree of dispersion of pigment within the cell is altered (fig. 20).   When the dark pigments are concentrated, the animal is pale; when they are dispersed, the animal is darker. The method by which this concentration and dispersion is obtained

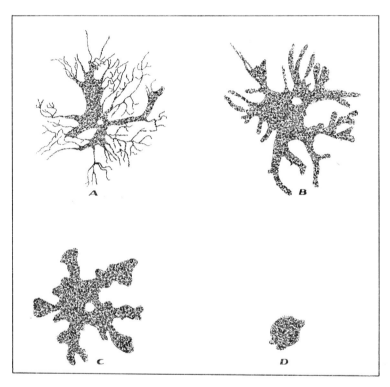

FIG. 20.   The possible degrees of pigment dispersion in a single chromatophore.   *A*. Pigment maximally dispersed.   *B*. and *C*. Intermediate conditions.   *D*. Pigment maximally concentrated. (From Kleinholtz.[298])

is still uncertain, but one possibility originally put forward to describe pigment movement in fish chromatophores may be applicable to Crustacea also.   This theory suggests that the pigment movements are associated with sol-gel transformations of protein molecules within the chromatophores.   When the protein is in a sol state, the pigment is dispersed ; when it is gelated, the pigment is dragged centrally and concentrated.[362]

The physiological responses can be divided into two groups, primary and secondary.[183]

(1) *Primary responses* are movements of pigment dependent on the intensity of illumination of the animal but independent of the eyes. Such responses are due to the direct reaction of the chromatophores to the intensity of the light incident upon them. Thus in *Uca*[57] increase in light intensity is followed by dispersal of both black and white pigments whilst in *Palaemonetes* the white pigment is dispersed.[59]   Primary responses are particularly obvious in young stages, but have also been observed in adult *Hippolyte*, *Eupagurus*, *Palaemon*, *Leander* and *Ligia*.

(2) *Secondary responses* are movements of pigment adapting the animal to its background. These are mediated exclusively via the eye. Where the requirements of the primary and secondary responses conflict, the latter are usually dominant in animals with the eyes intact. Thus individuals on a white background will have their dark pigment contracted even in bright light.

Background responses can be modified, however, by changes in temperature, since chromatophores apparently also play a part in thermoregulation as well as camouflage. Darkly pigmented animals will tend to absorb more heat than pale ones and it has been observed.[57] that, as an apparent counter to this, the dark pigment of *Uca* tends to be concentrated when the temperature of the environment is raised above about 15°C.

Two other factors, both cyclical effects, can also impose modifications on normal background responses. These are (1) colour changes associated with stages in the moulting cycle and (2) colour changes associated with diurnal and tidal rhythms.

(1) Immediately before and after each larval moult the amount of melanin-dispersing hormone in the eyestalk of the crab *Sesarma reticulatum* decreases and then increases again during intermoult. Similarly *Palaemon* apparently loses the capacity to bring about a maximal concentration of its red and yellow pigments for a time after moulting.[103]

(2) Colour changes associated with diurnal and tidal activities have been particularly studied on a fiddler crab, *Uca*, though also known to be present in other forms such as *Callinectes*.

Crabs showing diurnal rhythms darken by day and pale by night, the amplitude of the effect tending to be marked at high temperatures. A rhythm of colour change persists even if the animals are kept in constant darkness, indicating that the rhythm is endogenous. The period of the rhythm under constant conditions is not affected by temperature change.[61]

Tidal rhythms in *Uca* are characterized by increased dispersion of pigment some one to three hours after low water. Like the diurnal changes, the tidal cycle is endogeneous, persisting in constant conditions for a number of days at least. Indeed, crabs shipped by air from Woods Hole on the Eastern seaboard of the United States of America to California continued to match tide-associated colour variation of individuals left at Woods Hole.[62]

In all types of secondary response pigment movement is controlled exclusively by hormones, there being no nerve supply to the chromatophores. As in the case of the distal eye pigments, the degree of dispersion of the pigments depends on the balance maintained in the blood between antagonistic hormones. Thus, it is found that, if blood from the crab *Uca* is perfused through an isolated limb, the effect on the chromatophores depends on whether the donor of the blood had previously had fully expanded or fully contracted melanophores. When a limb is isolated the pigment tends to contract slowly, but blood from dark animals slows down the process while blood from pale animals accelerates it. The accepted explanation of this is that two different hormones are present, one normally concentrating and the other dispersing the pigment.[182] Diurnal and tidal rhythms are presumed to be associated with variations in the amounts of hormone released into the blood at different times.

The sinus gland of the eyestalk and post-commissural organ appear to be the main sites for the release of chromatophoretic hormones, though the hormonal systems involved do not appear to be constant in all Crustacea. A few examples of variations of melanophore response will illustrate this point. (1) Removal of the eyestalks of decapods, other than the crabs, causes the animals to darken. By contrast, the same operation results in brachyurans blanching. (2) Injection of sinus gland extracts from any decapod into shrimps (*Crangon*) lacking eyestalks brings about paling of the body. The sinus glands of *Carcinus*, *Pagurus*, *Libinia* and *Uca* apparently lack the principle which causes paling of the tail chromatophores of *Crangon*, though it is present in the sinus gland of *Crangon* itself and *Palaemonetes*. (3) All decapod eye-stalk extracts seem to have the capacity to concentrate the pigment of eyestalkless prawns and disperse that of eyestalkless crabs. Two separate hormones are believed to be responsible. Two possible explanations of these differences have been put forward: either (*a*) the sinus glands of *Crangon* and *Palaemonetes* contain three chromatophoretic factors whereas the glands of crabs contain only two; or (*b*) all sinus glands contain two principles one of which is different in crabs from that in *Crangon* and *Palaemonetes*.[183]

Comparable differences are found with respect to the regulation of red pigment cells. Thus, the most potent substance in the eyestalk of the crayfish *Cambarellus* results in dispersion of red pigment, whereas extracts of prawn eyestalks cause pigment contraction. To some extent such differences may be explained in terms of the comparative amounts of hormone stored in the eyestalk of different forms (and here one must constantly bear in mind that extracts will be most potent when much material is stored whereas in life a hormone, little of which is stored, may be equally active if its rate of production is high when required.) Thus, detailed study of the hormones controlling the red pigments of *Cambarellus* indicates that, although the most prominent hormone in the eyestalk is the one that disperses the red pigment, a principle that concentrates red pigment is also present.[182] Conversely, a factor that disperses red pigment has now been located in *Palaemonetes* eyestalk although the amount present is small.[60] Furthermore, red-dispersing and concentrating hormones can be extracted also from the supra-oesophageal ganglion and the circum-oesophageal connectives, though these are not the same substances as those in the eyestalk. Four different active materials affecting red pigment movement are therefore present at least in *Cambarellus*.[182]

The degree of dispersion of white pigment is also controlled by the eyestalk. Thus the sinus glands of *Orconectes immunis* have been shown to contain a substance which disperses white pigment.[56]

In summary, therefore, it may be said that at least the white, brown and red chromatophores are controlled in part by hormones from the eyestalk, though the precise response to injection of extracts differs according to the species involved, time of day, light intensity, temperature and previous history.

WATER METABOLISM

Removal of the eyestalks results in an abnormally large uptake of water and excessive swelling at the next moult.[90] This effect can be reversed if the eyestalkless animals are given injections of the ganglionaris X organ. The possibility has been noted, however, that the apparent action of eyestalk removal on the water balance may be an artifact due to the precipitated moult and not to any factor actively mediating the permeability. Thus the metabolic disturbance caused by eyestalk removal might conceivably result in the new cuticle having an abnormally high permeability or greater elasticity at the time of moult, independent of the action of any specific humoral agent. Since much of the water taken in by crabs at moult is drunk and absorbed from the gut, the possibility must also be considered that eyestalk removal results in a behavioural disturbance so that at the critical time when the cuticle is being expanded and hardened water is drunk faster than it can be eliminated. A further complication is raised by Hubschman's[266] observation that de-stalked *Palaemonetes* larvae are larger than controls at moult, even though the sinus gland is not functional at this time.

PROTEIN METABOLISM

Removal of the tips of the eyestalk slows protein breakdown, suggesting that there is a hormone there which stimulates protein catabolism. However, in *Eriocheir* the ratio of total nitrogen to body size remains constant at moult, even allowing for the increased water content in animals from which eyestalks have been removed. Removal of the sinus gland in *Hemigrapsus* results in a decrease in body protein and lipid content.

7. Pericardial Glands

The pericardial glands, which are composed of a plexus of nerve fibres in the lateral aspects of the pericardium, have been shown to be a source of hormones influencing the heart beat[367, 370] (fig. 16). This plexus, which is supplied by fibres running out from the visceral mass, lies across the openings of the first and second branchio-pericardial veins and so is well situated for the release of hormones into the blood as it enters the pericardium. Extracts of the plexus contain a material which increases the amplitude of contraction and rate of heart beat in *Cancer* and *Homarus*. It was at first thought that the agent responsible might be 5-6 dihydroxytryptamine.[91] However, this substance tends to break down rather rapidly in aqueous solution, whereas the rate at which pericardial extracts lose their activity is slower, indicating that if 5-6 dihydroxytryptamine is present it cannot be the only factor involved.[31] 5-hydroxytryptamine (5 HT) is detectable in the glands, but again it is doubtful if this can be solely responsible for the effect, since the heart of *Venus*, which is very sensitive to 5 HT, is not particularly sensitive to the pericardial gland extract.[372] Confirmation that some other active agent is present is given by the fact that treatment of the extract with trypsin largely abolishes its pharmacological effect on the crustacean heart, though this enzyme would not be expected to attack 5 HT.[107] Furthermore, hearts treated with an inhibitor of 5 HT show a normal response to gland extracts. The effect shown by trypsin might be expected if the effective agent of the gland extract was a polypeptide and Belamarich[31] has found evidence suggesting that there may be two peptides with cardio-accelerator properties in the extract. The position is further complicated, however, by the discovery that 6-hydroxytryptamine (6 HT) is also present in the pericardial glands, and that the heart is some ten times more sensitive to this substance than to 5 HT.[290]

The present state of our knowledge of these glands is therefore obviously incomplete, but before concluding that the evidence is so conflicting as to be useless it is worth remembering that the pericardial organs are ideally situated for the distribution of hormones in the general circulation and not merely to the heart. It is not unlikely, therefore, that several different active agents are produced.[369, 372] At least three types of neurosecretory cells contribute to the pericardial organ complex[369] (fig. 16). Two of these have their cell bodies in the ventral ganglion mass of the central nervous system and the cell bodies of the third type are in the pericardial organ itself.[369]   The size of the pericardial organ is larger in active crabs, such as *Callinectes* and *Ocypode*, than in sluggish ones, such as *Panopeus*.

An anterior ramification of the pericardial organ complex is situated close to the muscles of the maxillae.   This region is thus so placed that most of the blood passing over it will then flow over the scaphognathite muscles en route to the gill sinuses.   The nature of the secretion from this anterior ramification is not yet known, but the neurosecretory cells would seem to be well located for being able to influence either the activity of the scaphognathite or one of the gill functions.[370]

5 : The Blood System

The main body cavity of arthropods is a haemocoel, true coelomic spaces being restricted to the cavity of the excretory organ and genital ducts. Furthermore, unlike the situation in vertebrates, there is no development of an interstitial fluid and lymphatic system separate from the blood, as the blood itself comes into direct contact with the tissues. Since the blood thus combines the functions of interstitial and circulatory fluids, it is usually described as *haemolymph*.

In common with the blood of other animal groups, that of the Crustacea serves primarily as an equilibrating agent and as a vehicle for the transport about the body of food materials, respiratory gases, metabolic excretory products and hormones. Additional functions are found in some forms. The blood pressure affects the shape of the body in a number of species such as *Triops* (*Apus*) and *Artemia* which have thin cuticles and are of small body size. Less commonly, limb movements also are governed by blood pressure. Thus, in the ostracods *Gigantocypris* and *Doloria*, blood forced out of the trunk region by muscular action is used to provide the thrust which moves the mouthparts into a position suitable for the extrusion of the mandibular palps. Conversely, in *Apseudes talpa* one of the epipodites is driven in its functional stroke by muscular action, but the return stroke is effected by the blood hydrostatic pressure within the flexible peduncle.[136] These systems have obvious analogies with the pressure-operated extension of the limbs in araneid Arachnids, forms which lack extensor muscles in the limbs.

1. Haemolymph Volume

Various methods have been employed to determine the volume of the haemolymph: (1) by measurement of the dilution of a known amount of a substance injected into the blood, the substances most frequently used being inulin, thiocyanate and certain dyes; (2) by bleeding; (3) by determining the chloride space, i.e. the proportion of the total body fluid volume which would be occupied if all the chloride in the body were at the usual concentration in the blood. It should be appreciated that none of these methods is entirely reliable. Dilution methods suffer from the disadvantage that, following injection, some time must be allowed for complete mixing to occur before the blood is sampled. During this time some of the injected substance may be lost from the haemolymph by excretion, by absorption onto the surface of cells or by penetration into the intracellular compartment. If any of these effects is at all marked, the blood volume will be overestimated. Bleeding methods usually underestimate the volume, since it is difficult to extract all the blood. Excessive squeezing of the animal must be avoided, as this is liable to press fluid out of the cells. Determination of the chloride space, like the dilution methods, tends to overestimate the blood volume, since the assumption on which it is based—that all the chloride in the body is extracellular—is not strictly true.

It is clear therefore that the quoted values for haemolymph volume should be regarded as approximations rather than exact measurements. This is no great disadvantage, however, as it must be remembered (p. 74) that the blood volume even of individuals is in any case not constant but varies during the moulting cycle. The extracellular volume is at its maximum immediately after ecdysis and then gradually declines complementary to the growth of the cells during post-moult and inter-moult.

Considerable variations are found between the values of the haemolymph volume for different species. At one end of the scale the haemolymph volume of the branchiopod *Artemia*, indicated by the chloride space, is 87% of the total body water,[115] whilst at the other end, values for the volume in decapods lie in the range 25–37% of the body weight.[309, 381, 425] To some extent the range of volumes reflects differences in the functions

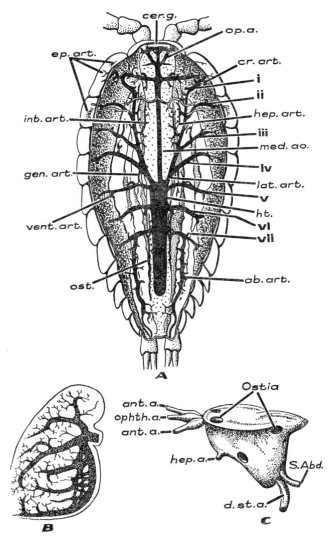

FIG. 21. A. The heart and arterial system of the isopod *Ligia*. B. Detail of one of the respiratory pleopods, showing the complex channelling of the blood. C. The heart of *Eupagurus*, showing the compact arrangement typical of decapod hearts and also showing the ostia. (A and B from Hewitt, C from Jackson in *Mem. Liverpool Mar. Biol. Comm.*)

(*a.* and *art.* arteries; *ao* aorta; *g.* ganglion); **i-vii** thoracic arteries; *ab.* abdominal; *ant.* antennary; *cer.* cerebral; *cr.* crural; *d.st.* descending sternal; *ep.* epimeral; *gen.* genital; *hep.* hepatic; *int.* intestinal; *lat.* lateral; *med.* median; *op.* and *ophth.* ophthalmic; *S.Abd.* superior abdominal; *vent.* ventral.

of the haemolymph in different species.   In *Artemia* the blood, in addition to its circulatory role, acts as a hydrostatic skeleton giving shape and mechanical rigidity to the body.[115]   This animal in fact " gives the impression of being an elastic sac . . . the actual tissues occupying a relatively small volume."[115]   In the larger Crustacea the skeletal function is taken over by the cuticle, and the haemolymph is restricted to a complex system of vessels and lacunae (fig. 21).

In some species a decline in the relative blood volume occurs with increasing size of the individual.   Thus *Procambarus clarkii* weighing 15 gm have a blood volume equivalent to about 30% of their body weight, while the volume is equivalent to only 10% of the weight in 40 gm animals.   A similar, though somewhat less, decline as size increases has also been noted in another crayfish, *Orconectes virilis*.[450]

Adjustments to the haemolymph volume can probably be brought about by varying the amount of fluid in the stomach, the urinary bladders[73] and, at least at the time of moult, pericardial sacs.[41, 144]

2. The Circulatory Pathways

A dorsally situated heart is present in most Crustacea though a true heart is lacking in the cirripedes and in many copepods and ostracods.   Where a true heart is absent, its function is often served by somatic muscles secondarily modified so that their contraction compresses a blood vessel.[86]   Alternatively, circulation of the body fluid may be brought about by movements of the body.   For example, in the copepod fish parasite *Caligus savala*, the posterior part of the body (abdomen and posterior thorax) is raised to an angle of 30°–40° with the remainder of the body and then lowered again at a rate of some 15 times a minute.   During the upward movement the posterior segments are compressed and the blood is driven anteriorly.   The abdominal vessels refill as the region relaxes.   The direction of blood flow is controlled by valves, so that blood tends to pass forward in a median sinus and posteriorly in lateral sinuses.[221]   Thus, even in the absence of a heart, an essentially one-way circulation of blood can be achieved. In a number of other forms lacking a heart, such as *Cyclops*,

Trebius and *Lernaeocera*, true circulation seems unlikely though peristaltic movements of the gut may provide a measure of mixing of the blood.[240, 492]

A heart, where present, may take various forms, ovoid in calanoid copepods, tubular in amphipods, isopods, and stomatopods, or rhomboidal in Brachyura. The wall of the heart is formed of an inner layer of striated muscle which is overlaid by connective tissue. The muscle cells, which anastomose to varying degrees, are usually arranged in a spiral or circular manner, though the compact hearts of decapods tend to have more complex muscle arrangements.

The walls of the single heart chamber are pierced at intervals (segmentally in tubular hearts) by ostia, the ports through which blood enters from the pericardial cavity to refill the heart after a contraction. Ligaments, taking their origin in the pericardium and ventral septum, serve to sling the heart within the pericardial cavity.

When the heart contracts, most of the contained blood is driven anteriorly (though some posteriorly directed flow may occur in forms with specialized abdominal arterial vessels) into the peripheral circulatory system. The complexity of the latter varies greatly. In most decapods, amphipods and isopods the circulatory system is largely composed of well defined vessels. Arteries lead from the heart, though these differ from their vertebrate equivalents in being composed only of endothelial lining and connective tissues. Muscles are absent except where there are valves. In some small forms, e.g. Branchiopoda, the main artery may open into the haemocoel without branching; but in Decapoda the arteries subdivide; and the thickness of the connective tissue progressively decreases, so that ultimately the finest vessels have only an endothelial lining and are thus comparable with the capillaries of vertebrates. Anastomosing capillary networks are sometimes built up from these vessels in association with the labyrinth and end-sac of the excretory organ and with parts of the central nervous system, though they are rare elsewhere. The capillaries ultimately open into lacunae, spaces ramifying between the tissues where there is little or no barrier separating the haemolymph and general body cells. Blood leaving lacunae is passed into sinuses which, although often large and ill-defined, differ from the lacunae in having an endothelial wall separating them from the general body tissues. An outer connective tissue layer is also sometimes present. In cases where the diameter of sinuses is such as to localize the blood flow they are sometimes termed veins.

The major circulatory route in decapods is as follows. Blood leaving the heart via the arteries is distributed to lacunae in various parts of the body. These in turn communicate with sinuses from which blood is taken to the ventral sinus which lies below the gut. From the ventral sinus it is passed into the gills and thence to the branchio-pericardial veins. These carry the blood to the pericardium, from which it enters the heart to complete the cycle. In the lobster, study of the fate of dyes injected into various parts of the body indicates that all blood returning to the heart must first have passed through the gills.[73]

In the smaller Crustacea there are usually fewer well-defined vessels and the lacunae are proportionately more extensive than in the larger forms.

3. The Heart Beat

When the heart muscles begin to contract, the ostia must be closed if blood is not to escape in the wrong direction. Closure is brought about by the combined effect of contraction of the cardiac muscles themselves and of the ostial muscles. The contraction of the former tends to elongate the ostia and so draw the lateral edges together. Complete closure is effected by the shortening of the ostial muscles which run round the rim of the ports. When muscular ostial valves are also present, as in the decapods, these are shut by the increasing internal pressure in the heart as the systolic contraction begins.

Valves, preventing backflow of blood into the heart, are present at the cardio-aortic junction. When the systolic pressure in the heart exceeds the pressure in the aorta, these valves open and blood is driven into the aorta.

The pericardial walls form an essentially rigid structure so that as the heart volume decreases during the expulsion of blood the pressure in the pericardium naturally falls. Blood is therefore drawn in from the branchio-pericardial veins. When the delivery stroke is completed the heart muscle relaxes, the cardio-aortic valves shut and the ostia open.   The elastic ligaments supporting the heart, which are stretched during contraction, then passively re-expand the heart while simultaneously haemolymph flows in through the ostia.   If the ligaments are cut, re-filling of the heart fails to occur.

4. The Heart Rate

The rate of heart beat is influenced by many factors, including body size, activity, temperature, respiratory stress, light and blood composition.   In some cases, however, change in heart rate following change in external conditions may be only transient as internal physiological adjustments are made and the heart rate is regulated.

BODY SIZE

In conformity with the general rule that heart rate varies inversely with body size, Crustacea of small size have faster heart rates than larger forms.   At constant temperature the relationship between heart rate and size can be expressed in the form[295]

Frequency/min $= 160 \times$ weight $(gm)^{-0.12}$.

This provides a reasonably good fit for many Malacostraca.   Not surprisingly, however, there are exceptions: the small prawn *Spirontocaris cranchi*, for example, average weight 0·39 gm, has been found to have about the same heart rate as *Leander serratus* of 3·2 gm when both are compared at similar temperatures.[200] The equation cannot be used to compare heart rate and body size in animals taken from different latitudes because of modifications of heart rate which occur as a result of temperature compensation, animals from a low latitude tending to have a lower heart rate at any given temperature than those from colder waters.   For example, *Spirontocaris* from Plymouth, England, have a heart rate 12% faster than those from the Mediterranean, if both are measured at 18°C.[200]

ACTIVITY

The heart rate naturally increases when individuals become more active, but Maloef[355] was unable to show that there is any simple relationship between metabolic rate and heart rate.

5. Control of the Heart Beat

In vertebrates the capacity to initiate the repetitive contraction of the heart is a property inherent in certain regions of the heart muscle. Such hearts are termed *myogenic*. By contrast, the heart of Crustacea, except the Branchiopoda, has to be stimulated to contract by nervous activity and is therefore *neurogenic*.

Two systems of nerves are of primary importance in the initiation and regulation of the heart beat, whilst additional nervous elements supplying the alary muscles and cardio-aortic valves play a supporting role.

Initiation of the heart contraction is dependent on the integrity of a system of nerve cells (cardiac ganglion) whose cells bodies and axons are restricted to the immediate proximity of the heart wall. If these cells are removed, the heart ceases to contract spontaneously. Further evidence of their excitatory function is provided by the fact that electrical activity can always be recorded from these cells a few milliseconds before contraction in the heart muscle, though at other times they are electrically quiescent.

The rate at which the ganglion cells initiate heart beats is controlled by a simple reflex which serves to co-ordinate contraction with the degree to which the heart has been expanded by the entry of blood during diastole This has been shown by perfusing isolated hearts with the cardiac ganglion intact. The rate of heart beat depends on the pressure of the perfusion fluid within he heart; the greater the pressure, the faster is the rate of beat until a maximum is reached. Conversely, isolated hearts may become completely inactive if drained of fluid. It is probable, therefore, that stretching of the dendrites of the ganglion cells is directly responsible for regulation of the frequency of discharge. In this respect the cardiac ganglion cells show some similarity to the sensory cells of the abdominal stretch receptors, which are likewise stimulated to activity by stretching of their dendrites.

Control over the rate of heart beat is also exercised by the central nervous system. In decapods this is mediated by three pairs of nerves, one pair being inhibitory and two pairs being accelerator fibres. The inhibitors originate ln the sub-oesophageal ganglion and the accelerators more posteriorly.[367]   At their distal ends these regulator fibres form synapses with the intrinsic neurones of the cardiac ganglion.

Experimental stimulation of the inhibitory fibres at a rate of 5–10 impulses per sec. decreases both the amplitude and the rate of heart beat. At high rates of stimulation the heart may even stop beating altogether. The mechanism of inhibition is comparable to that found in the inhibition of the sensory response of the abdominal stretch receptors. The arrival of an inhibitory fibre action potential at the synapse with a cardiac ganglion cell serves to repolarize the membrane of the latter, if it is already electrically active, and so blocks the initiation of nerve impulses which would normally excite the heart muscle. Treatment of the heart preparation with γ-amino-butyric acid mimics the effect of inhibitory fibre stimulation; so it is possible, though not proven, that this substance is the transmitter at the inhibitor nerve endings.

Acceleration of the heart occurs after repetitive stimulation of the accelerator fibres, unless it is already beating at its maximum rate.

Reflex variations of heart rate can result from stimulation of the body surface, particularly by pain or chemicals.[323] Slowing of the heart rate (bradycardia) can sometimes be produced in crayfish by blowing jets of water at the base of the thoracic legs. Pinching a limb may cause arrest of the heart for one or two beats. Conversely, touching the sternum and pleura of the first and second abdominal segments may greatly increase the heart rate.

Both high CO_2 and low O_2 levels in the medium can result in cardiac slowing,[322, 323] and this effect may also follow the removal of $Carcinus$[280] and $Homarus$[73] from the water. Other chemical stimuli which produce heart slowing include 5% $NaCl$[323] and L-glutamic acid in the medium bathing the gills.

Hormonal control of heart beat by the pericardial glands is considered on p. 97.

6. Blood Pressure and Flow

If Indian ink is injected into a lobster after removal of the heart, it spreads rapidly to all parts of the haemocoel[73] thus indicating that there are few, if any, barriers in the form of

valves to guide the direction of flow.   It might thus be concluded
that it is possible for the blood to move in a rather irregular
manner between the various interconnecting sinuses of the haemocoel.   In   fact,   though   irregular   movements   have   been
observed during violent movements such as the flicking of the
abdomen, study of the passage of injected dyes indicates that in a
resting animal the blood is continuously passing in one direction
along specific pathways.[73]   In the absence of venous valves this
movement could be brought about only by pressure differences in
various parts of the blood system.   All the available evidence
suggests that these differences are produced primarily as a result
of heart action, though the blood flow through certain restricted
regions may be aided by additional muscular pumps, such as the
cor frontale of some decapods.   This structure consists of an
expansion of the median artery closely associated with somatic
muscles which can constrict the vessel.

As blood pressure measurements are rather difficult to obtain
on animals with an open circulatory system, reliable records are
somewhat rare.   At present, comprehensive data is available only
for the lobster.   When this animal is resting, the pressures in the
gill sinuses and pericardium, and diastolic intraventricular pressure, are all of the order of 1 mm Hg.   Contraction of the heart
raises the intraventricular pressure to 9–22 mm Hg, the most
frequent level being about 13 mm Hg.   The pulse pressure
which provides the circulatory thrust is thus of the order of 12 mm
Hg.   Immediately outside the heart the mean arterial pressure is
a little lower (11 mm Hg) and there is a further drop along the
arterial vessels, with the result that in the haemocoel (ventral
sinus) the pressure is down to 2–6 mm Hg[73] (9·6 mm Hg in
Carcinus).[106]   The operation of the scaphognathite creates a
negative pressure of 1–3 mm Hg inside the gill chamber and this
probably is partly responsible for the further pressure drop which
occurs in the gills.

Pressure in all parts of the system rises when the animals
become active but the systolic pressure at all times remains in
excess of that in the remainder of the circulatory pathways.[73]

Despite the low magnitude of the pulse pressure, circulation is
comparatively efficient because the peripheral resistance is also
low.   The heart output per beat in lobsters weighing about 500

gm is between 0·1 and 0·3 cc.   The minute volume at the average heart rate of 100 beats per min is thus 10–30 cc.   The volume of blood in animals of this size is about 75 cc; so complete turnover is effected every 2–8 minutes.[73]   Whilst this is slower than in mammals of comparable size, much of the difference is accounted for by the relatively smaller volume of the latter's blood.   Certainly the large volume and low pressure of the crustacean system is not to be regarded as being indicative of a sluggish and inefficient circulation.   Indeed, in the limited respect that the lobster heart uses less energy in pumping a given volume of blood (because the peripheral resistance against which it is working is low) than does the mammal, it may be regarded as being more efficient than the latter.[368]

7. Blood Clotting

The capacity of the blood to form clots is rather variable in Crustacea.   For example, the blood of *Panulirus* clots rapidly into a firm gel; but that of *Homarus* forms only a soft gel.   No true clot is formed in *Loxorhynchus* blood, but protein fibres separate and form a floating mass.[437]   The details of the mechanism of blood clotting have not yet been worked out, and it will suffice here to say that it does not involve a process identical with that of vertebrates, since the removal of calcium has no effect.

6 : Respiration and Metabolism

The term *respiration* should perhaps be restricted to describe the processes whereby oxygen is utilized at the cellular level, but it is usually taken to include also the preliminary mechanisms by which an adequate oxygen supply to the tissues is ensured. In the smallest Crustacea diffusion across the general body surface may be adequate to bring sufficient oxygen to the tissues, and circulatory organs may then be poorly developed. In larger forms an efficient circulation is essential if diffusion distances are to be kept short. For the largest forms, and those living in environments with a low oxygen-tension, respiratory pigments are necessary to supplement the oxygen-carrying capacity of the blood. It is usual for oxygen-exchange between medium and blood to occur largely at specially modified areas of the body surface, and the movements of one or more pairs of appendages ensure that a continuous current of medium is passed over this region. The subject of respiration can therefore be most conveniently divided into three aspects, (*a*) gaseous exchange at the body surface; (*b*) oxygen transport in the blood and (*c*) the metabolic processes leading to oxygen consumption at the tissue level.

1. Exchange at the Body Surface

Uptake Across the General Surface

In many of the smaller Crustacea, such as the Copepoda and most Ostracoda, specific respiratory organs are generally absent, and sufficient oxygen for the animals' requirements penetrates across the more permeable areas of the cuticle. Even in larger forms, oxygen diffusing across the general body surface may supplement that taken in at the gills.   For example, the recolouring of reduced aniline blue injected below the cuticle of the isopod *Ligia* indicates that oxygen can diffuse across the body wall, especially at the base of the legs and through the ventral surface.[491] Dependence on permeation through the general surface tends to decrease in forms of larger body size, though even in an animal as large as *Homarus* some 3% of the total oxygen uptake occurs at sites other than the gills.[507]

Oxygen diffuses more slowly across a dry cuticle than across a moist one, and the relative humidity can affect the rate of oxygen uptake.   Thus it has been found that less oxygen is absorbed by the isopod *Ligia* when the air is dry than when the humidity is high.[490]   As might be expected therefore, those isopods which are more fully adapted than *Ligia* to survive short-term exposure to low humidities are forms which have developed a low dependence on cuticular oxygen uptake.   This is shown by the fact that painting over the respiratory pleopods (abdominal appendages) of *Porcellio* and *Armadillidium* reduces the oxygen uptake to 34% and 26% of normal, whilst the same treatment lowers the uptake of the less specialized *Ligia* and *Oniscus* only by 50%.[490]   The output of carbon dioxide by isopods is little affected by blocking the pleopods and so it must diffuse out of the body at other sites.[490]

SPECIAL RESPIRATORY ORGANS

Specialization of some region of the body surface for gaseous exchange is of general occurrence.   Many Crustacea have epipodites on the thoracic legs which, because of their large surface area, thin cuticle and well developed blood supply, are assumed to act as gills.   Forms which lack such epipodites often have areas of cuticle whose silver staining properties suggest a fairly high permeability.   Such areas may occur in diverse regions of the body, the latero-ventral corners of the carapace in *Neomysis integer*,[428] the ventral surface of the abdomen in *Asellus aquaticus*,[264] and the shell fields in Branchiura.[279]   Isopods lack thoracic respiratory epipodites but in these forms the abdominal segmental limbs are modified as gills.   The more specialized terrestrial forms such as *Porcellio* and *Armadillidium* have branching air-filled pseudotrachea in some of their pleopods.   Such structures may be of importance in increasing the respiratory surface without at the same time proportionately enlarging the rate of water loss by diffusion from the gill surface.

Ventilation of the gills of isopods is achieved by moving the appendages themselves. In amphipods thoracic gills are present but aeration of these is indirect, a current of water being produced by the beating of the pleopods. In terrestrial amphipods, such as the Talitrids of the Indo-pacific region, the movements of pleopods would clearly be superfluous, and appropriately the appendages are markedly reduced in size.[269]

Aeration of the larger gills of many decapods involves more complex processes. The delicate plate-like or filamentous gills are enclosed by lateral extensions of the carapace, the branchiostegites.

The flow of water over the gills is produced by the beating of the scaphognathites, the flattened exopodites of the maxillae. In *Carcinus* the action of the scaphognathite draws water into the branchial chamber through openings anterior to the bases of the thoracic legs and in particular through the ' Milne Edwards ' openings anterior to the chelipeds (fig. 22). Up to 80% of the total flow is via the latter openings. Exhalent water passes out through paired orifices between the base of the antennae and the mouth. The gill areas served by the water entering at the bases of different limbs is as follows. Water entering via the Milne Edwards' openings irrigates the small anterior gills, $1-5$ and also gill 6, the most anterior of the larger gills. The remaining gills are served by the much smaller volume of fluid entering at the more posterior limb bases. Much of this water is passed straight into the interbranchial chamber lying between the body wall and the gills. From there it flows across the gill lamellae into the main part of the branchial chamber and so joins the exhalent stream. A lesser volume directly bathes the outer ventral part of gills $7-9$ (fig. 22). These various current streams serve to bathe the inner surface of the gills and the lower part of the outer surface but do not adequately irrigate the dorsal part of the outer surface. A current over this last region is provided by a periodic reversal of the scaphognathite beat which supplements the normal small backward flow.[14] Reversal is rhythmic but the frequency can be affected by external conditions. For some reason it tends to be more frequent about the time of high tide and also when the oxygen level in the water is low.   A high level of CO_2 in the medium inhibits reversal.[14]   These effects reflect the overall rate at which water is being pumped through the chamber, since the rate of reversal is directly related to the volume of water pumped.[15]

The reversed flow has also been claimed to serve the function of cleansing the gills of any material clogging them; but, whilst it

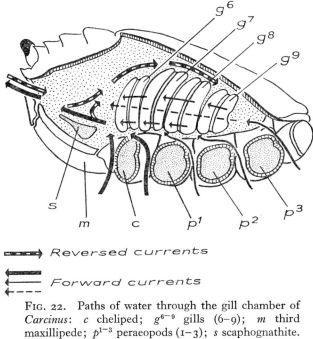

FIG. 22.   Paths of water through the gill chamber of *Carcinus*: *c* cheliped; g^{6-9} gills (6–9); *m* third maxillipede; p^{1-3} peraeopods (1–3); *s* scaphognathite. (From Arudspragasam and Naylor.[14])

may have an incidental effect in this respect, the absence of any maintained increase in reversal in water containing suspended particles seems to indicate that this cannot be its primary role.[14] Gill cleansing in fact is usually carried out by sweeping movements of the epipodites on the maxillipeds.

Carcinus, though not normally thought of as a burrowing crab, can readily dig itself below the surface if presented with a sandy substratum.   When it is buried, the entry of water at the base of its branchiostegites is restricted and the respiratory rhythm is modified, so that water is alternately taken in and expelled through the exhalent opening.[14]   It requires but a small adaptation from this condition to that found in the burrowing crab *Corystes*, where the predominant current is in the reversed direction.   This animal burrows itself almost completely in the sand leaving only part of its antennae exposed.   The inner borders of the antennae are fringed with long hairs so that when the two appendages are opposed, as they are when the animal is buried, they form a tube. Water is drawn down this tube by the reversed beating of the scaphognathites, passed over the gills and then out at the openings near the base of the legs.[217]

Ventilation Volume and Oxygen Utilization

The average rate at which large individuals of *Carcinus maenas*, *Cancer pagurus* and *Portunus puber* pump water at 16°C is about 1 cc/gm body-wt./min.   Small individuals of *Carcinus* average 1·5 cc/gm/min,[15] reflecting the higher metabolic rate of small animals.   Similar, though somewhat lower, rates have been found for *Homarus vulgarus* (0·3 cc/gm/min)[507] and *Procambarus simulans* (0·6 cc/gm/min).[321]

The proportion of oxygen in the respiratory current which is extracted by the animal (utilization) has been determined for only a few species.   In *Procambarus* it is 60-70%[321] and for *Homarus* and *Carcinus* 10-25%.[15]   The utilization is very variable, depending on the respiratory circumstances and species involved. It is usual, however, for it to be inversely proportional to the ventilation volume in any individual.   This is logical, since fast rates of flow allow less time for exchange to occur at the gills.   The various ways in which the rate of flow is adjusted to meet respiratory needs is considered later (pp. 157 ff.), but it is worth noting here that there is some evidence for the presence of a sensory receptor in the thorax of Crustacea which monitors the O_2 level and initiates changes in ventilation rate.[324]

2. Transport of Oxygen in the Blood

Oxygen passes through the cuticle and gill epithelium by diffusion.   Once in the blood, it is carried to the respiring tissues either wholly in solution, or partly in solution and partly bound to a blood pigment. Pigments are apparently absent from the majority of Crustacea, being known only from the larger species and from small forms which live in habitats liable to become deoxygenated.

Blood pigments serve to increase the oxygen-carrying capacity of the blood, since their properties are such that they combine with oxygen at the tension normally prevailing in the gills and release it at the tension found in the neighbourhood of active tissues. In some species the pigment, although having a high affinity for oxygen (becoming fully saturated with oxygen at a low oxygen-tension), is not fully saturated even in the blood leaving the gill vessels. In such forms, the pigment may have the secondary function of ensuring that a large gradient of tension is maintained across the gill surface, and so furthering a more rapid diffusion of oxygen than would otherwise be the case. The pigments also function in the transport of carbon dioxide and in the buffering of the blood.

Two respiratory pigments have been found in the Crustacea, haemocyanin and haemoglobin. In common with the pigments of most other invertebrates these occur in solution in the plasma.

HAEMOCYANIN

In the Crustacea the copper-containing pigment haemocyanin is not known to occur except in the decapods and stomatopods, though preliminary work on the spectral absorption curves of the blood of *Gammarus duebeni* suggests that it may also be present in some amphipods as well.

Haemocyanins are found also in the phylum Mollusca, but these differ from the crustacean blood pigments in having relatively more copper and less nitrogen in the molecule. In the latter group, the proportion of copper is about $0 \cdot 18\%$.[435]

Haemocyanins (HCY) are proteins of the globulin group and hence have a high molecular weight. The precise size of the effective unit of HCY is, however, somewhat uncertain. The minimum functional molecular weight for binding oxygen is about 74,000, which is a unit containing two copper atoms.[435] There is, however, some evidence to suggest that the normal stability of oxygen binding is achieved only when the unit contains eight copper atoms.   A unit of this size, which is able to bind four molecules of oxygen, would thus have a molecular weight in the region of 296,000.[435]   Sedimentation rate studies indicate that in the haemolymph units may be associated to form still larger units having molecular weights in the range 600–800 thousand.   As the HCY is present in the haemolymph only in solution, large molecular weights have the obvious advantage of restricting the loss of pigment in the excretory system, since smaller molecules, such as that of elasmobranch haemoglobin, can be filtered from the blood by the green glands of *Homarus*.

In the oxygenated state the pigment is blue, but it becomes colourless when deoxygenated.   In the oxygenated form it has characteristic spectral absorption bands at 275 μ and 340 μ in the ultraviolet and about 570 μ in the visible wavelengths.   The last band disappears on deoxygenation.   Oxygen is probably linked to HCY in the form of perhydroxyl free radical ion.[302] As it requires two copper ions for the carriage of each molecule of oxygen, it seems likely that there is a fluctuating valency state of the copper, half being in the Cu^{++} state and half as Cu^{+}.

The speed with which oxygen can be taken up or liberated by the pigment is comparable with that shown by mammalian haemoglobin: the time for half-loading of *Maia* HCY at 22°C and a pH of 8·6 being only 25 msec.[376]   By contrast, however, the oxygen-carrying capacity of haemocyanin containing bloods is low compared with those where haemoglobin is the respiratory pigment.   Values for oxygen saturation levels in the blood of ten marine decapods lie in the range 0·24–2·80 cc per 100 cc haemolymph, whereas in some vertebrate bloods 20 volumes per cent, or even more, may be carried.   As the oxygen level in the crustacean blood is thus at most only a little above that which would be in air-equilibrated sea water at atmospheric pressure and 15°C (0·5 vols %), the respiratory role of haemocyanin might perhaps be questioned. Any doubts must be dispelled, however, by measurements which show that the actual pressure of oxygen within the blood system is very low and hence that the amount of oxygen carried in physical solution is extremely small.   As a consequence, almost all the oxygen in the blood is in combination with haemocyanin.   Thus if *Panulirus* blood is equilibrated with air *in vitro* at $25\cdot4°C$ it contains $1\cdot99$ cc/%O_2 of which $1\cdot54$ cc/% (77%) is in combination with HCY. *In vivo*, by contrast, over 96% of the oxygen carried in the post-branchial blood is linked to HCY.[437]

The amount of oxygen carried by HCY is determined by a number of factors, of which the most important are the oxygen tension, the unloading tension of the particular haemocyanin, the temperature, and pH.

When other factors are held constant, the amount of oxygen carried by haemocyanin varies with the oxygen pressure up to a level at which the pigment is fully saturated (fig. 23*A*). In most animals the property of the respiratory pigment is such that it tends towards full saturation at the oxygen tension in the blood flowing past the respiratory surface, but is not capable of carrying so much oxygen at the tension pertaining in contact with respiring tissues. Consequently the pigment takes up oxygen at the respiratory organs and releases it to tissues with an oxygen demand. To make the maximum effect, a pigment should be such that it is approaching oxygen saturation in the respiratory organ but is largely unloaded at the tension in the neighbourhood of tissues during periods of peak oxygen demand.

Very few measurements have been made of the oxygen-carrying capacity of crustacean bloods *in vivo*; but such information as there is seems to indicate that the haemocyanin is not invariably fully saturated in the gills of decapods. In *Panulirus* for instance the HCY in the post-branchial vessels contains only a little over half the oxygen carried by the air-equilibrated blood *in vitro* (fig. 23*A*). The HCY clearly fulfils the function of a respiratory pigment, since the amount of oxygen in the pre-branchial vessels is only 22% of the saturation value[437]; but it would appear that a considerable part of the potential oxygen-carrying capacity of the blood is not being used. The crab *Loxorhynchus* shows a similar effect, the post-branchial blood being 68% saturated and pre-branchial blood 30% saturated with oxygen. It seems improbable that this incomplete loading of the haemocyanin in the gills of these forms is due to the pigment being poorly adapted to the needs of the animals in relation to the rate of diffusion of oxygen across the gill epithelium and the circulation rate. A possible explanation is that the full carrying capacity of the HCY is required only at times of special stress. Redmond[437] has suggested that the pigment may become more fully saturated at the gills at low temperatures, and that this would perhaps be advantageous if low temperature had differential effects on general metabolism and circulatory rates. There has as yet been no attempt to verify this idea experimentally, but it is known that the relation between loading tension and oxygen carrying capacity of HCY does vary with temperature. Thus in *Panulirus* the blood carries 50% of the saturation level of

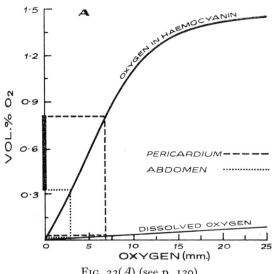

FIG. 23(*A*) (see p. 120)

oxygen with a partial pressure of oxygen of 14 mm at 25°C, 9 mm at 20°C, 6·5 mm at 15°C and 5 mm at 10°C[437] (fig. 23*B*). For any given loading tension, the blood will therefore bind more oxygen at a low than at a high temperature. The HCY of the crab *Loxorhynchus grandis* behaves similarly, but is even more sensitive to temperature change than is that of *Panulirus* (fig. 23*B*).[437]

The loading tension giving half-saturation with oxygen is of the same order of magnitude in other Crustacea.

Listing the half-saturation levels with greater precision would be valueless unless other conditions are precisely defined, since the degree of oxygen binding at a given tension and temperature may also be influenced by the pH and blood ionic composition.

In vitro experiments on *Homarus* blood indicate that, as in the case of haemoglobin, the more acidic the blood is made, the further the curve defining the relationship between oxygen binding and

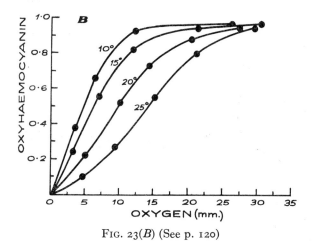

FIG. 23(*B*) (See p. 120)

oxygen tension is shifted towards the right (fig. 23C_1).   However, at a pH of about 6·0–6·5 (below the physiological level for this animal), there is a reversal of the effect, so that the half-loading

TABLE 13

Species	Temp. °C	½ Saturation mm O_2	Reference
Panulirus interruptus	25	14	437
Cancer irroratus	23	circa 12	436
Cancer borealis	23	circa 12	436
Callinectes sapidus	23	circa 12	436
Panulirus interruptus	15	6·5	437
Maia squinado	15	circa 14	493
Palinurus vulgaris	15	circa 14	493
Cancer pagurus	15	circa 14	493
Homarus vulgaris	15	circa 14	493
Procambarus simulans	25	3·5	325

tension moves lower as the pH is further decreased (fig. 23C_2).   The bloods of *Palinurus*, *Maia* and *Cancer* behave similarly.[258]

9

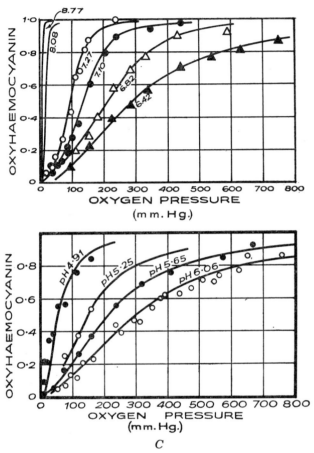

C

FIG. 23. The effect of various factors on the binding of oxygen by haemocyanin. *A*. The binding of oxygen at different tensions and the amount of oxygen actually carried in the blood of *Panulirus interruptus* at 15°C. The thickened portion of the ordinate indicates the volume of oxygen given up by haemocyanin as the oxygen tension drops from the level prevailing in the pericardium to that in the abdominal sinus. *B*. The effect of temperature on the oxygen dissociation curve of the haemocyanin of *Panulirus*. *C*. The effect of pH on the oxygen dissociation curves of the blood of the lobster *Homarus americanus*. (*A* and *B* from Redmond,[437] *C* from Redfield.[435])

120

It is doubtful, however, if this reversal is of any physiological significance in the normal temperature range in animals such as *Homarus, Panulirus* and *Loxorhynchus* since the difference in the amount of CO_2 carried in the pre- and post-branchial blood and the related difference in pH (0·02) is too small to exert any appreciable effect on the loading-tension curve of the HCY,[437] let alone depress the pH to a level where reversal would occur.

TABLE 14 (from Redmond[437])

	Post-branchial	Pre-branchial	V-A
Panulirus	10·09	10·62	0·53
Loxorhynchus	18·6	18·8	0·18
Homarus	5·19	6·04	0·85

At high temperatures, however, it is conceivable that pH changes in the blood might be sufficient to cause some shift in the position of the loading-tension curve, since *in vitro* experiments indicate that temperature and pH interact in their influence on HCY. The higher the temperature, the greater is the amplification of the shift caused by pH changes; conversely, the higher the pH, the smaller are the shifts due to temperature change.[437]

The level of HCY in the blood, like that of haemoglobin, is not necessarily maintained constant during the life of an animal. An extreme example of change in concentration of HCY is found in *Maia*. In this animal the HCY level in the blood builds up during the pre-moult stages and then declines rapidly after moult, until by the C_4 stage none is detectable.[567]

HAEMOGLOBIN

Haemoglobin is absent from the Malacostraca but occurs in at least a few species of most of the other crustacean groups, having been recorded in the Notostraca (*Artemia salina*,[438] *Branchippus*[209]), Concostraca (*Triops cancriformis*[203]), Anostraca and Cladocera,[318] parasitic Copepoda (*Mytilicola intestinalis*[209]), parasitic Cirripedia (*Septosaccus cuenoti*[410]) and Branchiura (*Dolops ranarum*).

In the blood, the pigment is in solution and not in corpuscles as in the vertebrates. It is not, however, restricted entirely to the blood, since in *Daphnia* it occurs also in the nervous system, muscles and eggs.

The small size of the crustacean species having haemoglobin has limited the extent of experiments on its properties and detailed studies have largely been confined to *Daphnia*. The haemoglobin from this genus consists of two fractions, the larger fraction having a molecular weight of 422,000 and the smaller a molecular weight of 34,000.[498b] *In vitro*, the affinity of the pigment for oxygen is high, the oxygen tension required for half-saturation in the absence of CO_2 being only 2·0 at 10°C and 3·1 at 17°C.[201] The presence of CO_2 decreases the affinity for oxygen. Thus 1% CO_2 causes a shift in the tension required to half-saturate the pigment at 10°C from 2 to 3 mm O_2.[201] Such a Bohr shift probably assists in the unloading of oxygen from the pigment in the neighbourhood of respiring tissues. Specific differences are found in the binding power of haemoglobins. *Ceriodaphnia laticaudata*, which tends to live in more polluted water than *Daphnia*, appropriately has blood with an even higher affinity for oxygen than that of the latter, the tension for half-loading being only 0·8 mm O_2 at 17°C.[201] Polluted waters often contain considerable amounts of carbon dioxide and in such circumstances a pigment showing a marked Bohr shift would be disadvantageous, because it would be unable to take up its full capacity of oxygen from the medium. Appropriately, *Ceriodaphnia* haemoglobin has only a very small Bohr shift in the presence of 1% CO_2.[201] Although the haemoglobins of *Daphnia* and *Ceriodaphnia* require such low tensions for half-saturation *in vitro*, it appears that the oxygen tension needed in the medium if the pigment is to be maintained in the oxyhaemoglobin form *in vivo* is very markedly higher. Spectral absorption bands appropriate to oxyhaemoglobin can no longer be detected at external oxygen tensions below 28 mm O_2.[201] A considerable gradient of oxygen tension must therefore be maintained across the cuticle if the blood pigment is to take up oxygen.

The amount of haemoglobin in the blood of *Daphnia pulex*, *D. magna*, *D. obtusata* and *D. hyalina* tends to vary inversely with the oxygen content of the water in which they are living.[202, 210] Presumably in this way the animals are able to maintain the total amount of oxygen transported by the blood to the tissues even when the external oxygen tension is below the level required to saturate the pigment fully. The variation in the amount of haemoglobin is particularly striking in the case of *D. hyalina*. This species is a member of the superficial plankton of lakes and, when in well oxygenated water, is colourless and contains mere traces of haemoglobin. When kept for a time in water with a low oxygen content, however, it develops enough haemoglobin to colour the animal pink. *Triops (Apus) cancriformis* can similarly increase its haemoglobin content in response to oxygen deficiency,[203] suggesting that this faculty may be common to all forms able to synthesize haemoglobin at all. It may be recalled that mammals have the same capacity; mountaineers acclimatized to high altitudes have a higher concentration of haemoglobin in the blood than individuals from lower levels. Whereas man's ability to increase his haemoglobin level is limited to a 20% rise, however, *Daphnia* may increase its haemoglobin concentration by as much as ten times.

As might be expected, individuals of *Daphnia* with a high haemoglobin content have several advantages over those with a low haemoglobin level, when both are placed in a medium with a low oxygen tension. The ones with much haemoglobin then gather more food, produce more eggs and survive longer than those with only little haemoglobin.[209]

The synthesis of haemoglobin depends on the nutritional state of the animals, a certain minimal amount of food intake being required to maintain the maximal amount of haemoglobin typical for a given oxygen concentration in the water.[210] However, even when the food supply and the oxygen tension are maintained at steady levels, the haemoglobin concentration does not remain constant but shows regular cycles of fluctuation. These fluctuations have been correlated with the reproductive cycle.[146, 210] A few hours before the moult, which occurs immediately before egg-laying, haemoglobin from the blood is transferred to the ovaries. After egg-laying the eggs contain haemoglobin and the level of haemoglobin in the blood has dropped to about two-thirds of its normal value. The concentration in the blood is built up during the time the young are developing in the brood pouch, and the cycle is then repeated with the release of the young and production of a new batch of parthenogenetic eggs.

Raising the temperature of the medium to 28°C results in an increase in the haemoglobin concentration in the blood, even if the oxygen concentration in the water is kept constant. This increase may be presumed to be in response to the higher utilization rate when the metabolic activity is raised and to the lesser affinity of haemoglobin for oxygen at the higher temperature, though a contributory factor may be the lower rate of egg-production at high temperatures.[206]

The addition of iron, particularly ferrous iron, increases the production of haemoglobin by animals in oxygen-deficient water.[206]

Re-aeration of water in which *Daphnia* with a high haemoglobin content are living results in a decrease in the haemoglobin level.[206]

Curiously, a high carbon dioxide level which, like high temperatures, would be expected to lessen the oxygen-carrying capacity of the blood, does not result in any increase in the rate of haemoglobin synthesis.[206] Possibly this may be related to the low sensitivity of crustacean respiratory systems to CO_2. For example, the respiratory rate of *Homarus* and *Austropotamobius* is not affected by CO_2 as long as the pH remains reasonably near the normal level.[205, 507]

Very high levels of CO_2, such as are obtained by bubbling the gas through water, will rapidly narcotize Crustacea; and Beadle[27] has shown that some species, e.g. *Gammarus pulex*, do not survive if high levels of CO_2 are maintained in the medium for more than about fifteen minutes. In the case of this animal, it has been shown that CO_2 narcosis is associated with a major leakage of potassium from the cells, the blood potassium level rising some 70% in ten minutes.[336]

Measurements of the CO_2 levels *in vivo* indicate that there is relatively little difference between the pre- and the post-branchial blood with respect to the amount of gas carried.[437]

3. Respiratory and General Metabolism

If the situation in *Hemigrapsus nudus*[374] is general to all other forms, the stored carbohydrate of Crustacea is largely in the form of glycogen in the integument, muscle and hepatopancreas, together with some sugars in the blood and also some sulphated muco-polysaccharide in the integument and hepatopancreas.

The nitrogenous polysaccharide, chitin, is present in considerable quantities in the integument, but this substance cannot be regarded as being available to the general metabolic pool once it has been laid down.

The concentration of glycogen in vertebrate muscle is on average about $0 \cdot 7 - 1 \cdot 0\%$ of the wet weight and the concentration in the whole body (wet weight) of *Hemigrapsus* is comparable, though there is a sex difference, the females containing almost twice as much as the males.[384]   The concentration in the hepato-pancreas of *Hemigrapsus* is on average $0 \cdot 76\%$[374] of the wet weight, as compared with the human liver level of about 5%.   As the level is low $(0 \cdot 3 - 1 \cdot 2)$ in *Cancer magister* also,[374] it seems likely that the crustacean hepatopancreas may normally store less glycogen than its vertebrate equivalent.   Naturally, as in the case of the liver, the glycogen content of the hepatopancreas is not constant but varies with the diet.   A high carbohydrate intake, such as is produced by the ingestion of glucose, may raise the concentration in the hepatopancreas to $1 \cdot 0\%$ in *Hemigrapsus*.[374] The concentration also fluctuates at different stages of the moult-ing cycle, being highest in the hepatopancreas of *Cancer pagurus* in late pre-moult, and falling rapidly at moult and post-moult to reach a minimum during the B_2 stage.[439]   Indeed, this effect is so striking in *Panulirus* that the hepatopancreas may be completely denuded of glycogen by the end of the B stages.[511]

The first metabolic reserves to be called on during starvation in mammals are those of the carbohydrate pool; and only when these substances are substantially depleted is there any appreciable utilization of fat and protein.   Starvation of *Hemigrapsus*, however, even when continued for twenty-three days, does not have any appreciable effect on the total glycogen levels of males, though there is about a 30% drop in females.[384]   Lipid stores also are unaffected by starvation for this period, but protein is depleted in females; which suggests that this group of substances may be more readily utilized as substrates than in mammals.[384]

Further indication of the relatively lesser role of carbohydrate as a metabolic reserve is provided by the low levels present in some marine planktonic forms.   In euphausids the whole-body wet weight carbohydrate concentration may be as little as 10% of that in *Hemigrapsus*, $0 \cdot 05 - 0 \cdot 08\%$ in *Meganyctiphanes*, $0 \cdot 09 - 0 \cdot 11\%$

in *Thysanoessa* and 0·08–0·17% in *Nematoscelis*.[432]   In other planktonic forms, such as the copepod *Calanus finmarchicus* (0·11– 0·40%) and the mysid *Neomysis integer* (0·19–0·23%),[432, 433, 434] it is closer to the *Hemigrapsus* level.

Respiratory measurements on these planktonic forms indicate that carbohydrate is not the only substrate being metabolized during short periods of starvation.[432]   A shift away from a purely

TABLE 15

Body composition of Hemigrapsus nudus *as influenced by starvation* (from Neiland and Scheer[384])

| | Normal | | Starved | |
	Male	Female	Male	Female
Bodyweight (gm)	10·5	9·3	9·5	7·4
Glycogen (mgm/gm)	0·69	1·24	0·8	0·8
Protein (mgN/gm)	12·71	14·96	12·2	12·08
Lipid, fat index/gm	4·31	6·13	4·76	6·40
Chitin, mg glucose/gm	3·91	4·23	3·89	4·09

carbohydrate metabolism appears to occur very early in the embryological life of *Carcinus* also, since there is a change in the respiratory quotient from 1·0 in the early cleavage stages to 0·72 when the yolk is still about four-fifths of the diameter of the egg, and to 0·83 at the time of hatching (fig. 24).[383]

In view of the apparent greater emphasis on protein metabolism in marine forms, study of carbohydrate metabolism in terrestrial Crustacea would doubtless repay attention, especially since the general suppression of nitrogen metabolism in terrestrial isopods[145] suggests that protein is unlikely to act as a major metabolic substrate in these animals.

Some of the above considerations *appear* to lead to the conclusion that carbohydrate metabolism may be of less importance in aquatic Crustacea than in vertebrates.   Indeed, at one time it was thought that the tissue metabolism of Crustacea might differ from that of vertebrates, because it seemed that labelled glucose injected into the lobsters *Panulirus japonica* and *P. penicillatus* did not give rise to labelled carbon dioxide.[462]   However, as glucose injected into the blood disappears too rapidly to be accounted for by the rate of carbon dioxide production or glucose secretion,[247, 462] it is apparent that at least a mechanism is present for the metabolism and storage of glucose. The earlier views on the inability of glucose to act as a respiratory substrate were later revised, following the observation that *Cancer magister* and *Hemigrapsus nudus* can both oxidize glucose to CO_2.[265, 374]

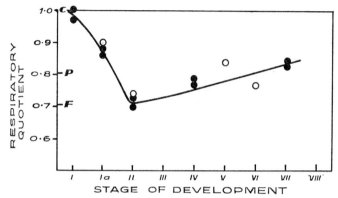

FIG. 24. Respiratory quotients of *Carcinus* embryos at different stages of development at 37°C. and at 15°C. *C*, *P* and *F* are the theoretical quotients for combustion of carbohydrate, protein and fat. (From Needham.[383])

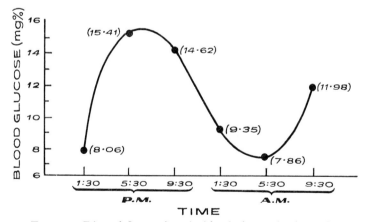

FIG. 25. Diurnal fluctuations in blood glucose in the crab, *Uca pugilator*. The figures in brackets give the mean glucose concentration for batches of 10 animals. (From Dean and Vernberg.[124])

At present a number of lines of evidence suggest that glucose plays quite as important a role in the metabolism of Crustacea as in that of vertebrates.

Glucose constitutes about 20%, or a little less, of the blood sugars of *Hemigrapsus nudus* and *Cancer magister*[351, 374]; but,

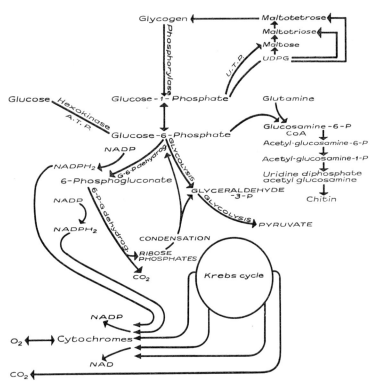

FIG. 26. Metabolic pathways in crustacean metabolism. (Based on information in the References 29, 348, 374, 411, and 532.)

whereas other major carbohydrate constituents of decapod blood, such as the oligosaccharides, maltose, maltotriose and malto-tetrose[351, 374] vary rather widely in concentration according to diet and time of year, the glucose concentration is maintained at a more constant level (fig. 25).[474] Diurnal fluctuations in the blood glucose of *Uca* can possibly be related to activity rhythms which are also cyclical.[124]   Furthermore, though a hexokinase for the conversion of glucose to glucose-6-phosphate has not yet been located,[374] metabolic pathways leading from the latter to glycogen formation,[374, 532] glycolysis via the Embden-Meyerhof pathway and hexosemonophosphate shunt[348] are well established (fig. 26). That the complete biochemical apparatus for carbohydrate metabolism is available thus seems reasonably certain.   Finally, in barnacles and the wood-boring isopod *Limnoria* carbohydrates are the first substrate to be utilized during starvation.[21, 218b]

GLYCOGEN FORMATION

In vitro the enzyme phosphorylase can catalyse either the synthesis or the breakdown of glycogen[110]; but it is certainly not the only, nor probably even the main, means by which polysaccharide is synthesized *in vivo* in either vertebrates[411] or Crustacea.[374, 532]   After injection of ^{14}C-labelled glucose into *Cancer magister* the ^{14}C label can subsequently be detected successively in the glucose-6-phosphate, maltose, maltotriose and maltotetrose components of the blood sugar.   Later it appears in the glycogen and mucopolysaccharide of the liver.[374]   This suggests that, as in the vertebrates, a pathway involving glucosyl units is utilized in the formation of glycogen.   That maltose and its polymers are involved is further indicated by two other points.   The injection of ^{14}C-labelled maltose results in the appearance of the label in glycogen more rapidly than does the injection of labelled glucose.[374] Again, whereas much of the label from maltose appears as $^{14}CO_2$ when maltose alone is given, the presence of added unlabelled glucose prevents this.[374]   Incidentally, this also provides additional evidence for the role of glucose as a normal respiratory substrate.

In vertebrates, an early stage in the production of glycogen involves the synthesis of uridine-diphosphate-glucose (UDPG) from glucose-6-phosphate.   The enzyme, UDPG-glycogen transglucosylase, which is involved in the transfer and polymerization of the glycosyls to form glycogen, has been found in the muscles, hypodermis and hepatopancreas of the crabs *Cancer magister* and *Hemigrapsus nudus* and the crayfish *Astacus cambarus*.[532]   The amount in the hepatopancreas is apparently lower than that in the muscles and hypodermis, but this may possibly be due to inactivation by the proteolytic enzymes of the former tissue during the separation.[532]

The activity of this pathway leading to glycogen synthesis is controlled by a factor from the eyestalk. Extracts of the eyestalk of *Hemigrapsus* inhibit the action of the transglucosylase enzyme from *Cancer magister* leg muscle and also, incidentally, the corresponding transglucosylases from a frog, toad, fish, gasteropod and lamellibranch.[531, 532] Similar inhibition of the enzyme is given by extracts of *Astacus* eyestalk. The eyestalk inhibitor is only partially deactivated by boiling; which indicates that the hormone is unlikely to be a protein.

The degree of inhibitory activity exerted by the eyestalk extract varies with the stage of the moulting cycle.[532] Extracts taken from animals in the C_4 and D_1 stages inhibit the transglucosylase, but extracts made from individuals in D_4 give a slight activation of the enzyme activity[532]; which is in good agreement with the cyclical nature of glycogen storage already noted.

It is known that eyestalk extracts contain a diabetogenic factor which raises the level of sugar in the blood when injected into normal animals,[1, 300, 462] but removal of the eyestalks causes depletion of the blood sugar in *Panulirus*[462] and *Orconectes*,[350] though not apparently in *Callinectes*[1] or *Libinia*.[300] The action of the transglucosylase inhibitor is compatible with that of the diabetogenic factor; for, if glycogen formation is suppressed, the blood sugar levels might reasonably be expected to rise. As yet, however, there is no direct evidence to indicate that hyperglycaemia and suppression of transglycosylase activity are actually produced by a single substance.

Other factors affecting the level of blood sugar include feeding and asphyxia. The blood sugar levels rise in *Astacus*[246] and *Libinia*[300] after feeding, and it appears that this rise is not directly associated with eyestalk fractions since it also occurs in eyestalkless animals.[300] Conversely, starvation depresses the blood sugar levels in *Carcinus*.[196]

Asphyxia normally induces hyperglycaemia,[300, 497] but does not do so if the eyestalks are removed.[300]

As might be expected from the observation mentioned above that the activity of the eyestalk inhibitor of transglycosylase varies with the phase of moult, there are comparable fluctuations in the blood sugar levels. *Callinectes sapidus, Austropotamobius pallipes* and *Maia squinado* all show raised blood sugar levels prior to moult.[26, 120, 147] The level of reducing sugar, calculated as glucose, rises from a mean value of about 50 mg/l in inter-moult *Carcinus* to around 200 mg/l in pre-moult animals.[456] The blood volume is calculated to increase to some 240% of the initial value at moult,[456] and this rise is sufficient to ensure that the dilution of glucose does not bring its blood level below the normal concentration. This is probably of considerable importance when the cellular requirements for respiratory substrates at this critical time are taken into account. One may be justified therefore in considering the mobilization of blood sugar prior to moult as being partly in preparation for the dilution of the body fluids.

CHITIN SYNTHESIS

The decrease in hepatopancreatic and integumentary glycogen stores that occurs during the pre-moult stages subsequent to D_2 and during the post-moult stages cannot be attributed solely to the need to raise the blood sugar levels. The decrease somewhat precedes the main period of formation of the new cuticle, since injections of ^{14}C-glucose result in labelling of chitin only if made in the D_4, A, and B stages[374]; but it is probable that much of the ultimate drop in hepatopancreatic glycogen can be attributed to the formation of new chitin.[439, 462, 523] The reason for the delay between the drop in glycogen level in the hepatopancreas and the formation of new chitin is that there is an intervening phase of intensive storage of glycogen in the epidermal cells under the new cuticle.[510]

Details of the biochemical stages in the synthesis of chitin have not been worked out for Crustacea; but it has been inferred that the route may have its starting point at glucose-6-phosphate and pass via acetyl-glucosamine as in fungi (fig. 26).[374] One of the final stages in the synthetic chain, uridine diphosphate acetyl-glucosamine, has been located in insect pupae[87] but has not yet been sought in Crustacea.

GLYCOLYSIS AND OXIDATIVE METABOLISM

It is only recently that clear evidence of the presence in Crustacea of the classical Embden-Meyerhof pathway of glycolysis has become available. The triose-phosphate of this process was shown to be present by Hu,[265] but he could not demonstrate either fructose-1-phosphate or 3-phosphoglyceric acid. However, the respiration of the hepatopancreatic tissue of *Orconectes* has now been shown to be sensitive to fluoride and iodo-acetate,[349] two of the known inhibitors of this pathway. Inhibition of the respiration of isolated hepatopancreas by sodium azide and cyanide suggests that the final stage in coupling to oxygen involves a cytochrome oxidase,[349] whilst the inhibitory action of antimycin further suggests that electron transport may involve cytochrome C.[349] There is further evidence for the presence of cytochromes a_1, a_3, b, c and c_1 in the hepatopancreas of *Carcinus* and the muscles of *Callinectes* and *Homarus*. Flavo-proteins, NADP and $NADPH_2$ have also been located.[29, 30, 505] Lobster and crab muscles have rather a low cytochrome b level as compared with the levels of a, a_3, c and c_1; and this is of some comparative interest.[505]

Cytochromes a and a_3 are present in relatively large amounts in *Callinectes* and *Homarus*, and it seems possible that these may subserve the function of a store of oxygen-equivalent material in the muscles, since these forms seem to have no muscle pigment store like the myoglobin of vertebrates.[505]

The tricarboxylic cycle (Krebs cycle) is present [349, 378] and is associated with the mitochondria.[378]

It is tempting, therefore, to conclude that glycolysis and oxidative metabolism are similar in Crustacea and vertebrates. It must be remembered however that, even though similar pathways can be demonstrated in different animals, this does not necessarily imply that their quantitative usage is similar. Even within a single individual crustacean the utilization of different pathways varies with the stage of the moulting cycle. Thus the respiration of *Orconectes* hepatopancreas is more sensitive to the glycolytic inhibitor, fluoride, during pre-moult stages than during inter-moult stages, indicating that the Embden-Meyerhof pathway is not the only one involved in glucose catabolism in the latter moult phase.[349]

The alternative glycolytic pathway of *Orconectes* hepatopancreas is blocked by 5-bromouracil, respiration of the tissue decreasing by 38% in inter-moult animals but by only 13% in pre-moult ones.[349] This inhibitor is known to block the hexosemonophosphate shunt (alternative names: ribose pathway, ribose shunt, direct oxidative pathway, C_1 pathway) in vertebrate liver (fig. 27). Additional evidence that this pathway is indeed the alternative to Embden-Meyerhof glycolysis in inter-moult animals is given by the fact that provision of 6-phospho-glycerate (one of the early steps in the cycle) increases the oxygen consumption of inter-moult, but not of pre-moult, hepatopancreas.[348] Furthermore, the amount of $NADPH_2$ produced following administration of 6-phospho-glycerate varies with the phase of the

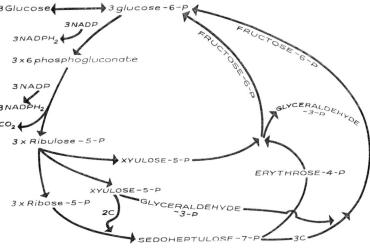

FIG. 27. The hexosemonophosphate shunt. (After Pesch and Topper.[411])

moult cycle in a manner predictably similar to the sensitivity to 5-bromouracil (fig. 28).[348]

Fig. 28 indicates that the hexosemonophosphate shunt provides a means whereby the Krebs cycle pathway can be by-passed. In vertebrates, it is thought that one of the main functions of the hexosemonophosphate pathway is to provide $NADPH_2$, which is then utilized in fatty acid and steroid synthesis.[411] This could perhaps provide the rationale for the cyclical utilization of this system in the Crustacea, since it is known[439, 460] that lipids are being laid down during inter-moult, reaching a maximum in D_1, whereas they are decreasing in amount during the pre- and post-moult stages, when the hexosemonophosphate shunt is inactive.

However, no direct proof of such a connection is available. The ribose sugars formed by the alternative pathway may be used also in nucleic acid synthesis.

Control over the activity of the hexosemonophosphate shunt is apparently mediated via an eyestalk hormone, since removal of

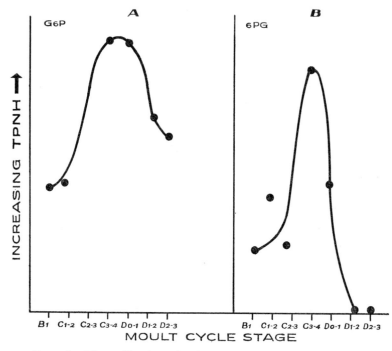

Fig. 28. The utilization of substrate by the hepatopancreas of crayfish during various stages of the intermoult cycle. Utilization is measured in terms of TPNH produced using (A) Glucose-6-P and (B) 6-phosphogluconate as substrates. (Modified from McWhinnie and Corkill.[348])

the eyestalks blocks the pathway. Extracts of eyestalk do not however have any effect on the rate of oxygen consumption of hepatopancreas *in vitro*,[349] so the hormone presumably does not affect the Embden-Meyerhof pathway.

From the fact that oxygen consumption is lower during inter-moult than at moult and that the hexosemonophosphate shunt is operative during inter-moult but not during pre-moult, it might be expected that the Embden-Meyerhof pathway activity would be reduced during inter-moult.   In fact the reverse is the case, the activity being greatest in inter-moult animals.[348]   A possible explanation of this paradox would be that Krebs cycle intermediates are being aminated to form amino-acids[348] for the protein synthesis that occurs at this time.   Alternatively, lipids may be being synthesized from acetate radicals.   In either case, the incomplete oxidation of the products of glycolysis would be compatible with the observed increase in glycolytic activity and decrease in oxygen consumption.

4. Factors affecting the Metabolic Rate

The metabolic rate is naturally influenced by the degree of voluntary activity; but both activity and/or the basal metabolic rate may also be affected by a variety of other factors.   Some of the more important of these include (1) environmental conditions such as temperature, salinity, CO_2 and oxygen pressures, (2) specific or race differences, (3) rhythmic effects associated with changing seasons, time of day or month, tidal state or phase of the moulting cycle, (4) body size and (5) nutritional state.

TEMPERATURE

There is no evidence to suggest that the metabolic rate of any poikilotherm is inherently insensitive to changes in ambient temperature, metabolic processes behaving like other chemical

TABLE 16

Oxygen consumption of Uca pugilator (*mm^3 $O_2/mgm/hr$*)
(from Kaplan, in Edwards[170])

15°C	20°C	25°C	30°C
0·034	0·052	0·077	0·100

reactions in this respect.[467]   For example the oxygen consumption of the crab, *Uca pugilator*, rises by a factor of three when the temperature is increased from 15° to 30°C.   Any rise in temperature thus tends to increase the metabolic tempo but it must also

10

result in a more rapid depletion of stored reserves when the animals are not feeding.   In apparent response to such effects many, but not all, Crustacea show a degree of homeostatic regulation of their metabolic rate.   Indeed, in the apt words of Sir Joseph Barcroft (1934), " Nature has learned to exploit the biochemical situation, so as to escape from the tyranny of a single application of the Arrhenius equation ".   Examples of the presence and absence of such homeostatic regulation of metabolic rate are shown by different species of the semi-terrestrial fiddler crabs, *Uca*.

Uca minax shows no sign of long term adaptation in its respiratory rate following a change in temperature.[503]   By contrast, *Uca pugnax* displays a considerable adaptation to temperature change over the next few days.   If a temperature change is made rapidly the subsequent Q_{10} (factor by which a process is altered for a 10°C change) for respiratory rate is 1·9, but if a slower change is made it is only 1·2 – 1·45.[503]   Full temperature acclimatization in *U. pugnax* takes about two weeks,[521] but when adaptation is complete this crab shows a much smaller variation in respiratory rate between 12° and 35°C than the non-adapting *U. minax* (fig. 29).

Fig. 29 also shows that there is a   considerable increase in the respiratory Q_{10} for *Uca pugnax* below 12°C.[503]   No such increase is apparent in the case of the non-adapting *U. minax*.   At temperatures below 12°C *U. pugnax* remains in its burrow, and it has been suggested that the increase in Q_{10} below this temperature will therefore confer two advantages: (1) Metabolic reserves will be conserved when the animal is unable to forage, (2) Only a comparatively small rise in temperature is sufficient to raise the metabolic rate to a level which enables the animal to undertake normal feeding activity.[503]   *U. minax*, lacking a high Q_{10} at low temperatures, remains inactive until the temperature is above 20°C,[503] possibly because only when such a temperature is reached is the metabolic rate high enough to enable normal activity.

Uca species make their burrows at or below the high tide mark, or in salt marshes; and it is interesting to note that those species which, by virtue of their temperature acclimatization, are able to be active at relatively low temperatures tend to replace *U. minax* on salt marshes, except where the influx of fresh water gives the more euryhaline *U. minax* an advantage.[503]

Seasonal changes in respiration are apparent in some species. In winter *Uca pugnax* use more oxygen at 7°, 17° and 28°C than do similarly treated summer animals.[520] The crabs used were from Florida where there is a considerable annual temperature range and where, consequently, it is advantageous to the animals

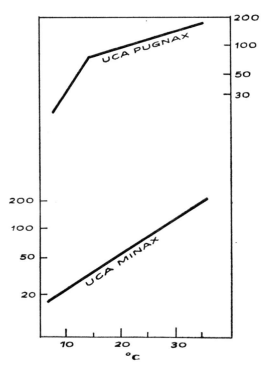

FIG. 29.   Semi-logarithmic plot of respiration (mm³ O₂/hr/gm fresh weight) against temperature, for *Uca pugnax* and *Uca minax*. (From Teal.[503])

to increase the metabolic rate at a given temperature when there is a seasonal decrease in temperature. *Uca rapax* from Jamaica, where the seasonal temperature fluctuation is less, show no sign of any such seasonal adaptation.[520] When races of *U. rapax* from two different latitudes are compared, however, they show differences in the relationship between temperature and metabolic rate.

Thus *U. rapax* from Florida have a metabolic rate at 7°C comparable with that of Jamaican individuals at 16°C.[520]

Latitudinal differences of this kind are very common in poikilotherms both at the level of the species within a genus and of the individual within a species. Forms of comparable body size from high latitudes in general have a higher metabolic rate at any given temperature than those from warmer waters though this is not invariably the case.[522b] As a result, the general level of activity of animals which spend their lives in cold waters is not dissimilar to that of their relatives from temperate or tropical areas. One of the more extreme examples of this is provided by the Antarctic amphipod *Orchomonella* which, though it lives permanently at a temperature of approximately—1·8°C, has almost the same rate of oxygen utilization in μl/gm body-weight as *Gammarus pulex* at 14°C.[13, 533] Similarly, a number of Arctic Crustacea from Kristineberg (Sweden) have a higher respiratory rate at any given temperature than related temperate forms from Plymouth (England), indicating the presence of a comparable temperature adaptation.[200]

TABLE 17

Comparison of the rate of respiratory movements of Arctic and temperate Crustacea at the same temperature (from Fox[200])

				Temp. °C	Respiratory movements per minute
Spirontocaris securiformis	A	*S. cranchi*	T	10	191 > 57
Pontophilus norvegicus	A	*P. spinosus*	T	10	145 = 141
Pandalus montagui	AT			11	260 > 161
Pandalus borealis	A	*P. montagui*	T	11	209 > 160
Apseudes spinosus	A	*A. latreillei*	T	8·5	91 > 54

A = Arctic
T = Temperate

The same type of latitudinal relationship occurs at the tissue level. Thus the heart rate of *Caprella acanthifera* from the Mediterranean is lower at 15°C than is that of individuals from Plymouth.[200]

It is still not clear to what extent the differences in metabolic rate of animals from various latitudes may be genotypic or phenotypic, though some results obtained on *Pachygrapsus crassipes* suggest a phenotypic effect.   This littoral crab shows no seasonal

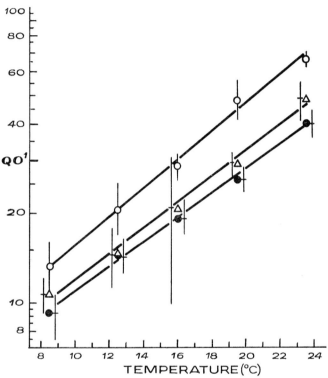

FIG. 30.   The response of QO' (a function of metabolic rate) mm³/gm/hr to change in temperature for groups of *Pachygrapsus crassipes* previously acclimatized to 8·5°C (○), 16°C (△) and 23·5°C (●) respectively.   (From Roberts.[453])

fluctuations in metabolic rate but its oxygen consumption depends on the temperature to which the animal has previously been acclimatized.   Diagrams of the respiratory rate plotted against temperature for crabs acclimatized for five weeks to a range of temperatures indicate that the oxygen consumption of individuals previously acclimatized to low temperatures is subsequently appreciably higher at all temperatures than those with a history of high temperature acclimatization [453] (fig. 30). A comparable shift is also found in the muscles, indicating, as does the latitudinal effect on heart rate, that the adaptation is at the tissue level rather than mediated solely by some central mechanism. The half time for temperature acclimatization in *Pachygrapsus* is about six days, though it varies somewhat according to the extent of the temperature shift; the larger the shift, the longer the adaptation period necessary.

TEMPERATURE-TOLERANCE AND LATITUDE

Latitudinal differences are apparent also in respect of temperature tolerance limits. At Plymouth the summer water temperature is about 15°C and individuals of the prawn *Pandalus montagui* taken there readily tolerate 17°C. Individuals of the same species from Kristiansund, where the summer temperature is only 7°C do not tolerate temperatures above 11°C. A similar effect is met with in regard to low temperature tolerance. *Uca rapax* from Jamaica are killed by exposure to 7°C whilst members of the same species from the Florida coast readily tolerate this temperature.[520]

The precise cause of heat death in Crustacea is unknown. The oxygen consumption of isolated gill, hepatopancreas, excretory organ, muscle and nerve taken from crayfish killed by high temperature are only slightly lower in the first hour after the animal's death than in tissues taken from a normal animal,[46] making it seem unlikely that a purely metabolic breakdown in any of these tissues is the cause of death. High temperatures do, however, result in one biochemical change; there is a marked shift in the ionic ratios maintained between blood and tissues. The blood sodium level falls and that in the tissues rises, whilst the opposite occurs in the case of potassium.[46] As neuromuscular responses, such as those of a crayfish claw preparation, function for a shorter period in a saline made to correspond with that of the blood of a heat-killed animal than in normal saline, it has been concluded[46] that heat death may be brought about in part by a disturbance of the normal ratios of blood sodium to tissue sodium, and of blood potassium to tissue potassium.

The Effect of Body Size

In general, the smaller individuals within a species, or a small-sized species within a genus, have a higher metabolic rate per unit weight and time than have larger individuals or species [520] (fig. 31). Thus the smaller species of *Uca* have higher metabolic rates than those of larger size, and the CO_2 production (which can be used as a measure of metabolic rate in other than early post-moult stages) declines with increasing size, e.g. in the wood-louse *Armadillidium pallasei*.[377]

Table 18

CO_2 *production by* Armadillidium pallasei *at* 21°C *in relation to body size.*[377]

	Weight, mgm				
	15	33	50	100	140
mm³ CO_2/gm/hr	200	174	144	112	94
mm³ CO_2/unit surface/hr	48·5	54·2	53	49·8	51·6

The possibility that changes in the surface area of the gills may impose limitations on the metabolic rate is suggested by the fact that with increasing size of animal there is generally a relative reduction in the area of gill per unit body-weight. Thus, in the spider crab *Libinia* the ratio of gill area to body-weight declines from about 700 mm²/gm body-weight in 50 gm animals to 400 mm²/gm body-weight in 600 gm animals. Similarly, *Menippe* has 1400 mm² of gill surface per gm body-weight in 20–30 gm individuals, but only 400 mm²/gm in 400 gm animals.[222]

If the area of surface across which gaseous exchange occurs is the limiting factor governing the metabolic rate, and if there is no differential growth of this surface relative to that of the rest of the body, then it would be expected that the metabolic rate (M) would vary as the 2/3 power of the body weight (W). Thus $M = \beta W^\alpha$ where β is a constant and the value of α should be 0·66 if this ' surface rule ' is applicable. A double logarithmic plot of metabolic rate against body size gives a straight line whose slope is α and which cuts the intercept at β. The figures obtained for the output of CO_2 by *Armadillidium* indicate that α in this case has a value of about 0·66. Similarly α for the uptake of oxygen by *Pachygrapsus*

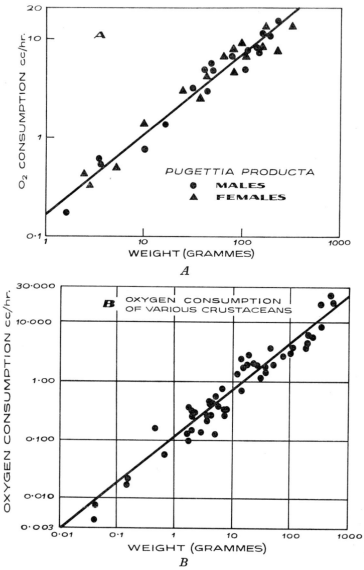

FIG. 31. The relationship between body size and oxygen consumption: *A*. in a single species (*Pugettia producta*), and *B*. in a variety of species. (From Weymouth, Crisman, Hall, Belding and Field.[549])

crassipes is 0·664[452] and by *Uca pugnax* is 0·621.[503] Other Crustacea found to obey the surface rule include *Daphnia pulex, Artemia salina, Asellus aquaticus, Oniscus asellus, Gammarus* sp. and *Astacus* sp.[32] In *Potamobius torrentium*, however, α approximates to 1·0 (45° slope) indicating that for this animal the metabolic

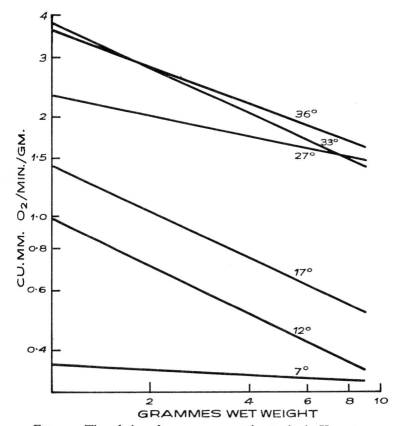

FIG. 32.   The relation of oxygen consumption to size in *Uca rapax* at different temperatures.   (From Vernberg.[521])

rate is directly proportional to the body weight.   Values for α intermediate between 0·6 and 1·0 occur in the crabs *Eurytium limosum, Sesarma reticulata, Uca pugilator* and *Pugettia producta* (fig. 31 *A, B*).[549]

A further complicating feature is that the value of α is not necessarily fixed for a given species but may vary with the temperature. Thus the O_2 consumption of *Uca rapax* from Jamaica shows less dependence on body size at 15°C than at higher or lower temperatures, whilst in an extreme case the O_2 consumption of *Uca rapax* from Florida is practically independent of body weight at 7°C though not at higher temperatures [520] (fig. 32). Such variations would seem to indicate that the limiting factor tending to govern metabolic rate is not the same under all conditions and that at low temperatures the rate of penetration of oxygen across the body surface is not limiting the metabolic rate.

Because of the differences in the value of α at different temperatures it is apparent that the Q_{10} for metabolism will tend to vary with the size of the animal tested,[520] an important consideration when comparative values are being assessed.

Another size-related complication in making comparisons between the metabolic rates of different species is that animals from high latitudes tend to be larger than those from warmer waters. Thus the maximum size of northern *Spirontocaris cranchi* and *Pandalus montagui* is greater than the southern races of the same species.[200] This size effect is probably phenotypic as Coker[106] has found that *Cyclops vernalis* reared at 29°C are smaller than those reared at 19°C, whilst the latter in turn are smaller than those raised at 9°C.

The Effect of the Moulting Cycle

There is a slow rise in the rate of oxygen consumption from inter-moult until ecdysis, followed by a decline during the post-moult stages.[36, 37, 181, 473] As the general locomotory activity is greatly reduced in the late pre-moult and early post-moult stages this rise in oxygen consumption may presumably be correlated with the increased catabolic and anabolic metabolism preceding and following moult. In *Cambarus immunis* the oxygen consumption rises to approximately twice the normal level for about five days before and after moult. As might be expected, the increase in metabolic rate is also apparent at the level of individual tissues, inter-moult hepatopancreas from *Orconectes* having a lower oxygen consumption than that of pre- and post-moult animals (fig. 33).[349]

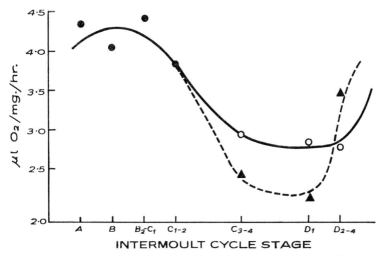

FIG. 33.   Endogenous respiratory levels of normal crayfish liver tissue during various stages of the intermoult cycle: ● Both pre-June and summer values; ○ Summer values; ▲ Pre-June values. (From McWhinnie and Kirchenburg.[349])

DIURNAL AND TIDAL CYCLES

Carcinas maenas, species of *Uca*, and presumably many other forms whose behaviour is influenced by the state of the tide, display both diurnal and tidal rhythms of metabolic activity. Distinct diurnal rhythms of oxygen consumption can be observed in *Uca pugilator* and *Uca pugnax* if the oxygen uptake for each hour of the day is averaged over a period of fifteen days.   In *U. pugnax* the rhythm is characterized by a maximum utilization of oxygen at about 8 to 9 a.m., a fall in O_2 consumption to a minimum at noon, followed by a second, smaller, peak at 10–11 p.m. and a second nadir near midnight.   This diurnal rhythm is apparently inherent since it continues over a 15-day period even when the animals are kept under constant conditions of illumination and near constant temperature.[58]

When the oxygen consumption of *U. pugnax* is considered on a daily basis (instead of taking the average for each hour over several days) and the values on successive days are compared,[58] it is apparent that the peaks and troughs of utilization shift in a systematic manner.   Peaks usually occur at about one hour before low tide on the shore where the animals were collected.   This is the time when the animals would be foraging away from their

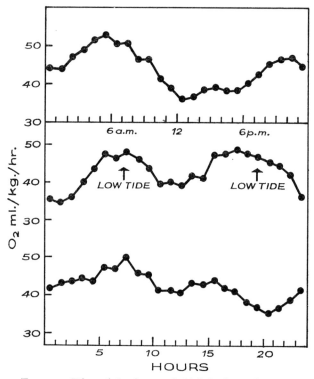

FIG. 34.   Diurnal rhythms and tidal rhythms of oxygen consumption in the crab *Uca pugnax*.   *Upper diagram*: the average diurnal rhythm; *Centre diagram*: the average tidal rhythm; *Lower diagram*: a new cycle of variation obtained by randomizing both the upper and centre diagrams, which shows the existence of a lunar cycle.   (From Brown, Bennett and Webb.[58])

burrows.   The minimum oxygen uptake occurs at the time of high tide, when *Uca* would normally be resting in its burrow (fig. 34).

The rise and fall in metabolic rate is therefore presumed to reflect changes in activity associated with the tidal cycle.   Like the diurnal rhythm, the tidal rhythm is maintained under constant conditions in the laboratory for at least fifteen days.[540]   Throughout this period the daily shift in the time of high tide is matched by a corresponding 50-minute shift in the peaks and troughs of oxygen uptake, so that the animals remain constantly in phase with the state of the tide at the place where they were collected.[58]

During the winter months, October to January, the tidal cycle is less obvious in *U. pugnax* and this correlates well with the fact that the animals are largely confined to their burrows during this period.[541]

EFFECTS OF HABITAT ON METABOLIC RATE

In crabs there is a tendency for the rate of oxygen consumption to be higher in the more terrestrial species,[18, 519] and this can be correlated with the generally greater activity of the semi-terrestrial forms as compared with that of their more aquatic relatives.   One has to run to catch an *Ocypode*, whereas a slow walk suffices for *Carcinus*!

TABLE 19

The relationship between habitat and oxygen consumption
(modified from Vernberg[519])

Habitat*	Species	Weight gm	O₂ µl/gm/min
A	*Libinia dubia*	123	0·42
A	*Callinectes sapidus*	140	1·14
InT	*Menippe mercenaria*	238	0·51
InT	*Panopeus herbstii*	11	0·93
St	*Uca minax*	6	1·28
St	*Uca pugilator*	2	2·03
T	*Sesarma cinerea*	1	2·21
T	*Ocypode albicans*	44	2·35

* A = Sub-tidal, InT = inter-tidal, St = Sub-terrestrial,
   T = Supra-tidal

In view of this increased uptake it might at first sight seem curious that there is an inverse relationship between oxygen uptake and gill area, the more terrestrial forms having reduced gill number, volume[409] and area[222] (table 20).   In aquatic species the gill filaments are supported and kept apart by water. When such forms are removed from water and the gill cavity drained, there is a tendency for the unsupported filaments to collapse against their neighbours and so restrict the area available for the uptake of oxygen. Suggestive evidence that such an effect may result in oxygen lack is given by the observation that the heart rate of *Carcinus* falls when the animal is removed from water just as it does if kept in water containing too little oxygen.[280] Reduction of the gill area in terrestrial crabs may help in reducing the danger

TABLE 20

Gill area (mm²/gm wt) of crabs from different habitats (from Gray[222])
Aquatic

Callinectes sapidus	1367	Portunus gibesii	1003
Areneus eriborus	1301	P. spinimanus	901
Ovalipes ocellatus	1288	Libinia dubia	748
Hepatus epheliticus	1099	L. emarginata	566

Low tide		Intertidal		Above tide	
Menippe		Uca pugnax	770	Sesarma cinerea	638
mercenaria	887	U. pugilator	624	Ocypode	
Panopeus herbstii	874	Sesarma		albicans	325
		reticulata	579		
		U. minax	513		

of collapse, whilst remaining commensurate with the greater O_2 uptake of these animals, since the faster diffusion rate of oxygen in air than water enables a high concentration to be maintained at the respiratory surface without the necessity of making extensive respiratory movements. The importance of the latter point seems to be illustrated by the greater O_2 uptake by amphibious crabs in water (table 21) than in air since it is likely that the difference is due to the need to make greater respiratory movements in the former medium.

Reduction in the number of gills occurs in the Ocypodidae, which have only 12 gills, as opposed to the 16 in aquatic crabs such as *Callinectes*. This, coupled with the decrease in area of the remaining gills, reduces the ratio of the surface of the gills to body weight to well below that found in other crabs. This is compensated for in *Ocypode* by the development of a vascularized accessory respiratory area in the gill chamber.

TABLE 21

Rate of oxygen uptake by amphibious crabs in water and in air
(from Teal[503])

	ml/kg/hr	
	Water rate	Air rate
	207	141
Uca pugnax	204	160
	186	195
Uca pugilator	170	87
	150	97
Uca minax	112	101
	144	134

Differences in metabolic rate are also observed in animals from various types of aquatic habitat. Two of the factors which have been related to oxygen consumption are salinity and degree of disturbance of the water.

THE EFFECT OF WATER MOVEMENT

Individuals of the isopod, *Asellus aquaticus*, taken from a fast-flowing stream have a rate of oxygen utilization about one and a half times that of individuals from a slow stream when both are measured in the laboratory under constant conditions.[211]

THE EFFECT OF SALINITY

As animals living in fresh water and dilute brackish water maintain their body fluid concentration above that of the medium, they have to do work to replace ions lost in the urine and by diffusion across the body surface. The higher metabolic rate of many fresh-water species by comparison with that of isotonic marine relatives and of individuals transferred from a high to a low salinity has been correlated with this fact. Thus the marine isopod *Idotea neglecta* has an oxygen uptake of only about one-third that of the fresh-water *Asellus aquaticus*[207] and the marine *Gammarus locusta* and *Marinogammarus marinus* have a lower oxygen uptake than the fresh-water *Gammarus pulex*. Euryhaline species such as *Carcinus maenas*,[465] *Gammarus chevreuxi*,[345] *Ocypode albicans*,[188] *Hemigrapsus oregonensis*[127] and *H. nudus*[127] all show an increase in oxygen consumption when transferred to a more dilute medium. Conversely, the metabolic rate of crayfish declines when they are transferred from fresh water to a medium isotonic with their blood. Results obtained with *Palaemonetes*, *Pachygrapsus* and *Artemia*, however, indicate that too facile an acceptance of these results as being indicative of the measure of osmotic work done is to be avoided.

Palaemonetes varians, as a form which maintains its blood hypotonic to sea water and more concentrated salines but hypertonic to media less concentrated than 25‰, might be expected to show its minimum oxygen uptake when isotonic with the medium if the amount of additional energy utilized in osmotic regulation is additive to a basal utilization rate. Prawns taken from a marsh pool with an average salinity close to isotonicity with the body fluids do indeed have their minimum oxygen uptake at this salinity. However, prawns of the same species whose normal environmental salinity is 1·3‰ show a minimum oxygen uptake at 6‰[344] even though the blood concentration of these animals is likely to differ only slightly from those in the more saline media.[391] Hence, this minimum oxygen uptake is not at the isotonic point. Furthermore, the rise in O_2 utilization between the minimum and extreme levels, some 600% (fig. 35) is so large as to make it improbable that it is due solely to osmoregulatory activity.[344] Similarly, careful studies of the effect of salinity and temperature on the oxygen consumption of *Hemigrapsus nudus*, *H. oregonensis*[126, 127] and *Uca*[227] also indicate that changes in oxygen uptake cannot be directly correlated with osmoregulatory energy expenditure.

A possible alternative explanation for the increased O_2 uptake by *Palaemonetes* placed in other than the environmental salinity, is that there is an increase in locomotor activity. Such an increase in activity has been noted in other forms. *Gammarus duebeni* becomes hyperactive when its medium is diluted, and *Pachygrapsus* makes efforts to escape when put in any medium other than sea water.[227]

The change in metabolic rate would in any case scarcely be expected to bear a direct relationship to the gradient of concentration maintained between an animal and its medium; for, with the exception of a few more or less homoiosmotic forms such as

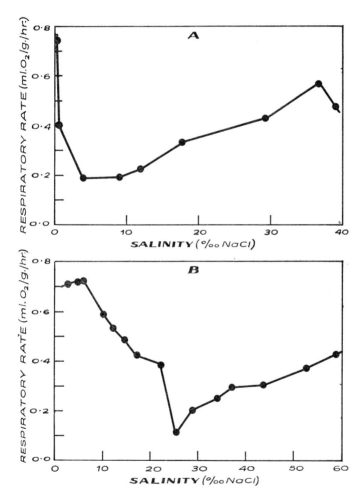

FIG. 35.   The respiratory rate of *Palaemonetes varians* at different salinities of external medium.   *A*. Animals from a low salinity habitat;   *B*. Animals from a high salinity habitat. (From Lofts.[344])

Palaemonetes varians, any change in the medium is reflected in changes in the gradient of ions between blood and cell as well as between blood and medium.   Consequently a change in medium concentration is likely also to affect metabolism at the tissue level, as well as the mechanisms responsible for regulation of the blood. A lowering of blood concentration will tend to result in a decreased sodium entry into the cells, since the gradient for sodium between blood and cell is decreased.   Hence it is probably that sodium has to be extruded less rapidly from cells when the blood is diluted and that the energy utilization in cellular ionic maintenance will also decline.[419]   It might therefore be theoretically possible for an animal with a relatively low surface permeability to balance any rise in metabolism needed to maintain an increased gradient between blood and medium by allowing a drop in blood concentration just sufficient to gain an equivalent metabolic saving at the cellular level.   In such a case, once the rise in metabolism associated with osmoregulatory adjustments to the amino-acid concentration in the cells was complete, no difference would be expected in the overall metabolic rate at different salinities.

At first sight *Artemia salina* might appear to approximate to this condition, since there is no difference in the oxygen consumptions of individuals reared at 35‰ and 140‰.[219]   However, despite the fact that the blood is very hypotonic to the medium at

TABLE 22 (from Gilchrist[219])

Salinity ‰	% females with eggs or embryos	Mean length of females, mm
115	72	8·42
140	49	7·85
160	15	6·88

this upper concentration there is nevertheless some rise in the blood level above that of animals in 35‰.[115]   An increase in the osmoregulatory work would therefore be expected both at the body surface and at the cellular level.   The fact that no such increase in metabolic work is observed suggests that energy may be diverted from other processes such as growth and reproduction. This may well be the case since females reared in concentrated brines are smaller and produce fewer offspring than those in more dilute media.[219] The size of individual animals can also be important in determining the magnitude of the alteration in oxygen consumption following a change of salinity. Small individuals of the crab *Sesarma plicatum* in 50% sea water have an O_2 consumption of only about half that when they are in tap water. By contrast, there is little difference in the O_2 consumption of adults in these two media.[352]

All in all it seems clear that interpretation of variations in metabolic rate following a change of salinity is rendered difficult by the variety of possible interfering effects and is not to be lightly undertaken.

SPECIES DIFFERENCES

As will already be apparent from table 19, there may be considerable differences in metabolic rate between species of the same size and from the same habitat. Very active aquatic forms such as the swimming crab *Callinectes* have higher basal metabolic rates than more sluggish forms such as *Libinia*. The active forms also have a larger gill area (table 20). Similar differences can be found between terrestrial species. Thus *Uca minax* has a lower metabolic rate than either *U. pugnax* or *U. pugilator*,[503] whilst the uptake of oxygen by *U. pugilator* is some 70% greater than that of *U. pugnax*[58] (table 21). Species differences are also found at the tissue level. Weight for weight, the mid-gut gland of *Callinectes* has about twice the O_2 consumption of that of *Libinia*,[519] and the oxygen utilization by the gills of crabs varies widely, the more active forms again showing the greatest levels.[519]

INANITION

In Crustacea, as in other animals, starvation decreases the metabolic rate (*Uca*,[520] *Astacus*[64]).

OXYGEN TENSION IN THE MEDIUM

Any animal living in an environment where the oxygen tension fluctuates, must either have a respiratory system capable of maintaining an oxygen uptake suitable for its requirements over a range of external tensions, or suffer limitations in its activity when a drop in tension decreases the rate of O_2 uptake.   Examples of both types have been found in Crustacea.   As perhaps might be expected, forms such as *Homarus*[11, 507] and *Callinectes*, which live in sub-littoral or inshore waters unlikely to show reduced oxygen levels, are not well adapted to oxygen lack, and their uptake is directly related to the external tension at all levels below that of

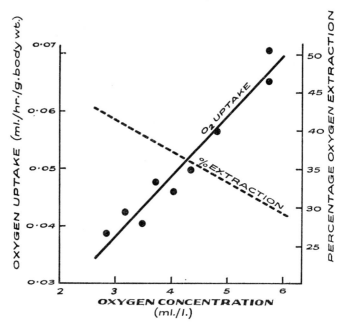

FIG. 36.   The effect of oxygen concentration of the medium on oxygen uptake, and the percentage of oxygen extracted by the gills of a lobster.   (From Thomas.[507])

atmospheric O_2 pressure (*circa* 150 mm Hg).   The crayfish *Procambarus clarkii* and *Cambarus bartoni* behave similarly [353] (fig. 36). Species in which the O_2 uptake is limited in this way by the external tension are termed *conformers*.   By contrast, many forms live in situations where there may be temporary depletion of oxygen; thus *Pugettia producta*,[549] often found in intertidal pools, *Orconectes virilis*[250] from slowly-flowing fresh water and *Chirocephalus diaphanus*[506] from temporary fresh-water ponds all show a measure of independence of the external O_2 tension. If the tension falls below atmospheric oxygen tension the rate of uptake by these ' *regulators* ' remains unaffected until a certain critical tension is reached. Below this critical tension, uptake is related to tension as in the conformers. The adaptive significance of the ability to regulate uptake despite fluctuating oxygen levels in the environment is obvious.

The critical oxygen tension (T_c) of the regulators does not have a fixed value but varies with species, body size and temperature. In *Pugettia producta* the T_c is at about 50–70 mm Hg O_2,[549] whilst in *Orconectes virilis* it is somewhat lower at 40 mm.[250] The effect of size is particularly well marked in *Orconectes immunis*, small individuals (4·3 gm) having their T_c at about 20% of atmospheric oxygen tension, whilst larger ones (9·1 and 17·1 gm) can regulate only down to 30% and 40% of atmospheric pressure respectively at 25°C.[245] As metabolic rate is influenced by temperature, a rise in temperature would be expected to exert an increased stress on the mechanisms responsible for maintaining an adequate and constant uptake of oxygen. Work on *O. immunis* suggests that this is indeed the case, since it fails to regulate so effectively when the temperature is raised. At 16°C, uptake is constant at tensions from atmospheric down to a critical level at about 60% atmospheric pressure. At 24° and 30°C, however, the relationship between oxygen consumption and tension more closely resembles that of a conformer than a regulator at tensions below 85% of the air saturation level [551] (fig. 37). There is thus no absolute distinction between regulators and conformers; and this is further exemplified by the observation that species such as *Procambarus clarkii* and *Cambarus bartoni*, which are conformers at normal temperatures and atmospheric pressure, regulate their O_2 consumption when presented with oxygen tensions higher than this.[353]

The precise physiological significance of the critical tension of regulators is difficult to evaluate. It has been suggested that the critical tension is that at which the blood pigment is just fully saturated.[437] This would not necessarily imply any direct relationship between the external O_2 level and the tension required to saturate the blood pigment *in vitro*, since the internal O_2 tension will ultimately be determined by the rate of tissue utilization and the rate of diffusion across the respiratory surfaces.   If the body surface acts as a partial barrier to the diffusion of oxygen, as seems to be the case in at least some crustacea,[325] then a fast rate of oxygen uptake can be achieved only if a considerable gradient

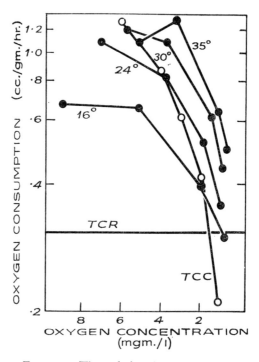

FIG. 37.   The relation between oxygen consumption of *Orconectes immunis* and oxygen concentration at different temperatures.   *TCR* indicates the theoretical curve of a regulator; *TCC* indicates the theoretical curve of a conformer.   (From Wiens and Armitage.[551])

is maintained across the respiratory surfaces. The tension of oxygen in the medium in contact with the respiratory surfaces will depend on that of the medium as a whole and the effectiveness with which the medium is replaced at the surface as the oxygen in it is utilized by the animal.   Its maximum value is obviously that of air-saturated water, or about 150 mm Hg O_2.   The tensions in the blood leaving the gills of *Panulirus interruptus*, *Loxorhynchus grandis* and *Homarus americanus* are, on average, only 7, 8 and 5·3 mm respectively.[437]   In fact the haemocyanin of these animals is not saturated, therefore, even in the immediately post-branchial blood and, as already noted, *Homarus* has no critical tension in the normal range of oxygen tensions in the medium.   The crayfish *Procambarus simulans*, which has haemocyanin which is fully saturated *in vitro* at a tension of only 15 mm, shows two breaks in its oxygen tension curve.[325]   Conceivably one of these might represent the level at which the blood is saturated.   It remains uncertain, however, whether the critical tension is always related to the saturation level of the blood pigment.

Insufficient time is available for equilibrium to be reached during gaseous exchange across the gills of decapods.   In *Homarus* the normal oxygen utilization is between 35% and 40% of that in the water flowing through the branchial chamber.   When the rate of flow is artificially depressed to one-sixth of its usual value, the utilization rises to 90% at 18°C and may be even higher at 8°C.[507]   Conversely, if the rate of flow is artificially increased the percentage of extraction falls.[507]   Despite the increased utilization when the rate of flow is lowered, the amount of oxygen taken in by the animal under these circumstances is decreased.   When the flow is one-sixth of normal, the uptake is only a half to a quarter the usual level.   It might be expected that an abnormally fast rate of respiratory flow would result in an increase in the total oxygen uptake, but in *Homarus* this does not happen.[507]   It is perhaps not surprising, therefore, that this animal does not increase the respiratory flow when the oxygen level in the medium is reduced. It does respond to lowered O_2 tensions, however, by increasing the percentage utilization (cf. fig. 36).[507]

In its normal environment *Homarus* is unlikely to experience low oxygen levels.   Freshwater species and intertidal forms such as *Carcinus* are more likely to meet such conditions and so be more fully adapted to them.

The freshwater *Procambarus simulans* responds to a low environmental oxygen tension by increasing the respiratory rate and ventilation volume.   As the utilization rate remains constant, however, over a wide range of O_2 tensions, the overall effect is that the O_2 consumption falls off as the tension declines.[325]   Thus, as in the lobster, the process which appears to be directed towards maintaining the level of oxygen uptake is inadequate when the external O_2 concentration falls.

The regulator *Carcinus* seems to combine the adaptive features of both the previously mentioned forms.   The ventilation rate rises if the oxygen tension falls, and remains at a high level as long as the O_2 content of the medium exceeds about 20% of the saturation level.   Accompanying this increased ventilation there is also a slight rise in the utilization rate, and together these factors enable the rate of oxygen uptake to be maintained at near the normal level over a range of oxygen tensions.[18]   In fully oxygenated water an increase in ventilation volume is accompanied by a decrease in utilization rate.[15]   Changes in the utilization rate are, therefore, presumably under physiological control, but how this is exercised is uncertain.   The most obvious means by which utilization could be regulated would be by (1) varying the area of gill exposed to the ventilation flow, (2) varying the permeability of the gill membranes to oxygen, or (3) varying the rate of blood flow through the branchial vessels.

It is possible that the blood system of *Carcinus* is better co-ordinated to respiratory requirements under conditions of low oxygen tension than seems to be the case in *Procambarus simulans*. A decrease in oxygen tension results in a slowing of the heart in this latter animal, and this, by making the circulation more sluggish, seems likely to decrease the oxygen uptake rate and so oppose any advantage which accrues from an increase in ventilation rate.[322]   However, as some preliminary results indicate that the heart rate in *Carcinus* also slows when exposed to low O_2 levels,[280] the role of the circulatory system in regulating utilization remains uncertain.   No evidence is yet available concerning the other two possible factors.   The fact that the basal metabolic rate can vary with the oxygen tension in some forms suggests that the metabolic rate is to some extent directly governed by the oxygen tension at the cellular level.   Evidence to show that the larger Crustacea can build up an oxygen debt is meagre; but there is an old suggestion that this may be possible in *Orconectes*.[250]   It seems improbable that marine forms such as *Callinectes* and *Homarus* could develop a major O_2 debt since, apart from rather large amounts of cytochromes a and a_3, these species have no intracellular pigment like myoglobin or oxygen equivalent substance.[505]

Homarus, although unlikely to be exposed to fluctuating oxygen tensions, will meet variations of temperature and, though it cannot respond adequately to the former, it responds to a rise in temperature by increasing both the ventilation rate and the percentage of oxygen extracted from the respiratory stream.   It thus matches the increased tissue requirements.[507]

It has already been noted that one effect of raising the temperature of the medium is to increase the oxygen requirement.   This, by stressing the capacity of the respiratory system to provide more oxygen, tends to make animals which show regulation of O_2 consumption at low temperatures into conformers at higher temperatures.   Comparison of the effects of temperature on the oxygen consumption of related species from different habitats reveals ways in which these responses might be involved in ecological adaptations.   *Orconectes immunis*, under conditions of high temperature and low O_2, retains its ability to regulate O_2 uptake somewhat better than does *O. nais*.   As the former is found principally in ditches, muddy pools and slow streams, all of which may be expected to have a higher maximum temperature and lower oxygen concentration than the shallow streams where *O. nais* is most common, this physiological difference is appropriate. *O. immunis*, size for size, has a lower metabolic rate than *O. nais*, and this factor, though advantageous in enabling it to regulate O_2 uptake in conditions of stress, may perhaps be important in preventing the colonization of the habitat occupied by the more active *O. nais*.[551]

The cave-dwelling *Cambarus setosus* survives for longer periods in an enclosed jar than epigean crayfish can, because like other subterranean animals, it has a very low metabolic rate and so uses up the available oxygen less rapidly.[70]   Similarly, pond species of crayfish tend to survive longer and reduce the oxygen levels in closed jars to a lower level than do stream species.[400, 401]

7 : The Neuromuscular System

The presence of a stout articulated cuticle provides a firm skeletal basis for the attachment of muscles and the arthropods alone of the invertebrate phyla have their somatic musculature solely in the form of discrete units. Typically, in each muscle the individual fibres run from an origin on the cuticle to an insertion on an apodeme (intucking of the cuticle) in a neighbouring segment. The form of the muscles varies somewhat with the function performed. Where the movement required at the insertion is large but no great power has to be exerted, the apodeme is short; and the muscle fibres may be of considerable length, several centimetres in the case of the larger decapods. Where power output is required, it is an advantage to have many fibres acting in parallel. A muscle design commensurate with this is achieved by the extension of the apodeme towards the segment containing the muscle. Many short fibres taking their origin over a considerable surface area can then be inserted on the enlarged apodemal surface.

The somatic muscle fibres resemble those of vertebrates in being multinucleate and, like the latter, are formed by the fusion of separate primordial cells. The fibres are striated and, in rapidly contracting muscles, the striations are closer together than in those which contract slowly.

The motor nerve supply to the muscles is more direct than in such phyla as the Echinodermata and Annelida, since there is no synapse intervening between the central nervous system and the muscle surface, as there is in these two groups.

The larger nerve axons have a Schwann cell sheath and myelin is often present in the Malacostraca. The Natantia have thick layers of myelin on some axons, and this may be interrupted at irregular intervals in a manner somewhat resembling the nodes of Ranvier in vertebrates.[259]   It may be presumed that the myelin sheath serves to increase the conduction velocity of fibres of small diameter since myelinated axons of 35 μ diameter from *Palaemon serratus* have about the same transmission rate as the 100–250 μ giant fibres of *Procambarus clarkii*, which have little myelin.[259]

The musculature of the Crustacea is controlled solely from the central nervous system, the peripheral nerve net which is present in such groups as the Annelida, Mollusca, Echinodermata and Platyhelminthes having been lost. Furthermore, the motor nerve supply from the central nervous system to the muscles in Arthropods is different from that of other animals, since there are only a few nerve axons.   In vertebrates the motor axons may be numbered in hundreds; in the Crustacea there are only four, or fewer to each limb muscle.   Perhaps the most extreme example of this reduction in motor axon supply is that of the opener muscle of the dactylopodite and the stretcher (extensor) of the propodite of the thoracic legs of decapods which share a single axon.

In addition to motor nerves, most muscles receive one or two axons which, when active, inhibit muscle contraction.   Such inhibitory fibres are apparently confined to the Arthropoda and perhaps to the Crustacea; for even in insects their presence is uncertain.

The complete motor and inhibitory supply to the distal segments of the thoracic limbs of several decapods has been carefully worked out.[557]   Minor differences occur between groups, particularly in respect of the inhibitory supply, but the brachyuran condition described below may be regarded as fairly typical of the general innervation. The seven most distal muscles, carpopodite extensor and flexor, accessory flexor of the carpopodite, extensor and flexor of the propodite and the abductor (opener) and adductor (closer) of the dactylopodite receive between them a total of 15 axons.   Of these, 12 are motor axons and 3 are inhibitory.   Three muscles, the closer of the dactylopodite, the flexor of the propodite and the extensor of the carpopodite receive 2 axons each.   Three more, the accessory flexor of the carpopodite, the opener of the dactylopodite and the extensor of the propodite receive 1 axon each which, in the case of the last two muscles, is a shared single axon. Finally, the carpopodite flexor has 4 motor axons. Five muscles share one of the inhibitor fibres; but, in addition to it, the opener of the dactylopodite and the extensor of the propodite each have a second inhibitor (fig. 38).[557] As these last two inhibitor fibres are separate, it is possible for the two muscles to have a measure of independence of action despite their sharing of a single motor axon.

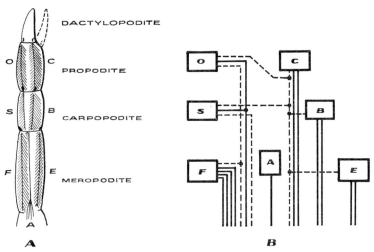

FIG. 38.   The innervation of the muscle of a crab leg. *A*. The seven most distal muscles of the limb. (From Hoyle and Wiersma[261]). *B*. The motor and inhibitory innervation of these muscles: *O* opener, *C* closer, *S* stretcher, *B* bender, *F* main flexor, *E* extensor, ——— motor axons, - - - - inhibitory axons. (From Wiersma and Ripley.[557])

The function of the inhibitor which supplies five muscles is not clear but it has been suggested that it may become active at the time of moult, to suppress muscular activity in the limbs until the new cuticle is sufficiently hardened to withstand this without distortion. Such peripheral inhibition, rather than the central inhibition of motor nerve activity, is presumed to be necessary because of the potential danger of accidental mechanical stimulation of the neuromuscular system during moult. Possible support for this idea comes from the observation that no activity can be detected in this common inhibitor in non-moulting animals.[77]

The Crustacea also differ from the vertebrates in the manner in which the nerves terminate on the individual fibres.   In vertebrates the axons branch within the muscle to supply a small number of fibres.   On each fibre the axon ends at a single end-plate, a specially modified part of the muscle fibre surface.   By contrast, the axon or axons supplying crustacean muscle, branch

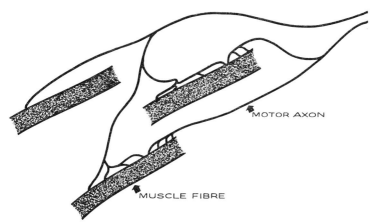

FIG. 39.   Diagrammatic representation of muscle innervation, to illustrate (1) that single axons can supply more than one muscle fibre, (2) that there are many nerve terminations of each axon on the muscle surface, (3) that all terminations tend to be on one side of the fibre, (4) that more than one axon may supply one muscle fibre.

repeatedly on the surface of the muscle and again on the surface of the individual fibres (fig. 39).   Each muscle fibre receives many axon terminations scattered along its length.   The endings, however, are simple and there are no end-plates of the vertebrate type, though the fibre membrane may be folded at the point of contact.[515]   Some insects do have end-plates comparable with those of vertebrates.[260]

Where a muscle is supplied by a single axon, it is obvious that this axon must innervate all the muscle fibres.   There is evidence that, in the case of muscles supplied by several axons, some fibres receive branches from more than one axon, though the individual

axons do not necessarily supply all fibres. Thus, in the carpopodite flexor of *Panulirus* 38% of the 421 fibres studied respond to
stimulation of one axon, 26% to two axons, 29% to three axons
and 7% to all four (Furshpan quoted by Hoyle and Wiersma[261]).

Differential innervation of the fibres can make it possible for
qualitative as well as quantitative differences to occur in muscle
response following stimulation of different nerve axons. In the
crabs *Dordanus asper* and *Dromidiopsis dormia* the flexor of the
propodite can rotate the propodite about its longitudinal axis
in addition to the normal bending movement. Activity in one of
the two axons supplying it causes rotation in one direction and
activity in the other rotates the joint in the opposite direction.[558]
The difference can be correlated with the distribution of the two
axons within the muscle since, though it appears that both axons
supply all fibres, the lateral parts of the muscle receive a more
profuse innervation by one than by the other.

1. The Response of Muscle to Stimulation

In mammals stimulation of a motor nerve results in an ' all or
nothing ' contraction by each of the muscle fibres it supplies.
Graded contraction of the muscle as a whole is determined by the
number of axons, and hence of motor units, active at any one time.
From the description of the limited number of axons supplying the
muscles it will already be obvious that the crustacean system does
not lend itself to the production of graded contractions by this
means. Nevertheless one does not need to watch any crustacean
long to see that it is capable of displaying both quick and slow
movements. In crayfish, graded movements are displayed in
such activities as walking, feeding and the swinging of the chelae
into a menacing position, fast movements in the snap of the chela
or flick of the abdomen in the escape reaction.

Investigation of the neuromuscular responses to stimulation
has largely been restricted to such muscles as the claw-closer in
decapods, which have two motor nerves. Early experiments indicated that stimulation of the nerve supply at different intensities but
constant frequency resulted in different types of muscle response.
Later it was realized that this was because the two motor axons
have different thresholds and, when active, cause different muscle

responses.   When the two axons are stimulated independently, a single stimulus to the larger of the two results in a twitch of the muscle.   A single stimulation of the other gives no muscular response, but repeated stimulation is followed by a slow contraction.[516]   From these differences in the mode of muscle response, the two axons are termed *fast* and *slow* respectively.   It must be recognized, however, that not all muscle fibres respond in an identical manner to activation by the two types of axon.   Four forms of muscle innervation are recognized[261]:

Group 1.   All muscle fibres respond to both fast and slow axon stimulus; but, whilst the response of some to the fast axon is greater than that to the slow, in other fibres the reverse is true. Examples, the closer of *Panulirus* and *Cancer*.

Group 2.   All the muscle fibres respond to both slow and fast axons.   The response to the slow axon is uniform in all fibres but that to the fast axon differs between fibres.   Example, closer of *Cambarus*.

Group 3.   The muscle is divided into two parts with the fibres of one part responding either to the slow axon or to the fast axon but not to both; whilst the other half is like a Group 2 muscle. Example, proximal part of the closer of *Pachygrapsus*.

Group 4.   Muscles in which all the fibres behave similarly but with marked differences in the response to slow and fast axon stimulation.   Example, extensor of *Panulirus* and closers of *Epialtis* and *Randallia*.

Even within the leg of a single individual, muscles occur with the characteristics of several of these groups.[261]

An even more extreme example of such differences in the properties of muscle fibres in response to stimulation is seen when the deep and superficial flexor muscle fibres of the abdomen of crayfish are compared.   The superficial fibres always exhibit smoothly graded contractions, i.e. they are tonic fibres.   The main function of these fibres is in the postural control of the abdomen where slow and maintained contraction is necessary.   By contrast, the deep flexor fibres contract rapidly on stimulation giving a twitch, i.e. these are phasic fibres.   They have not been observed to be capable of tonic contraction.   Tonic contraction is unnecessary, however, as their function is to produce the violent flexion of the tail in the escape reaction. [289b]

Clearly, given such a degree of morphological difference in axon distribution within muscles and physiological differences in the axon type and response by the fibres to axon stimulation, a great variety of types of contraction can be achieved. Many of these differences are associated with electrical events at the neuromuscular junction.

ELECTRICAL PROCESSES AT THE NEUROMUSCULAR JUNCTION

The inside of resting cells is at a lower (negative) potential than the outside (see p. 15 for a discussion of the reason for this). The potential difference (resting potential) in crustacean muscles is comparable in magnitude with that of the excitable tissues of other groups of animals, being in the range 55–75 mV. In any one fibre the value is maintained at a fairly constant level, though fibres in different muscles of an animal may have different resting potentials.[261]

On the arrival of a succession of nerve impulses at the neuro-muscular junction, local electrical events occur at most or all of the post-synaptic regions along the muscle fibre, and these may be followed by additional localized or more general changes in the membrane potential. Three main types of event are recognized, (1) junction potentials, (2) spike potentials, (3) maintained depolarization. All involve a temporary or more sustained depression of the membrane potential on a local or general basis. The magnitude of the change and the rate at which it occurs vary from fibre to fibre, but also depend on whether stimulation is via the slow or the fast axon.

THE ELECTRICAL RESPONSES FOLLOWING STIMULATION OF THE SLOW AXON

When a slow axon (or the single axon to the crayfish abductor dactylopodite) is active, small depolarizations (junction potentials) are produced at the muscle membrane in the region of the synapse after the arrival of each nerve impulse. From each of the many synapses on a fibre the electrical change spreads outwards. It is not propagated, but the decline in its magnitude is small. Even an artificial potential change produced at a single point near one end of a muscle fibre 2–3 mm long declines to only about 60% of its initial value by the time it reaches the other end.[390]   It is thus readily understandable that, when junction potentials are produced naturally at many points along the membrane, a practically uniform change in potential occurs over the whole membrane.

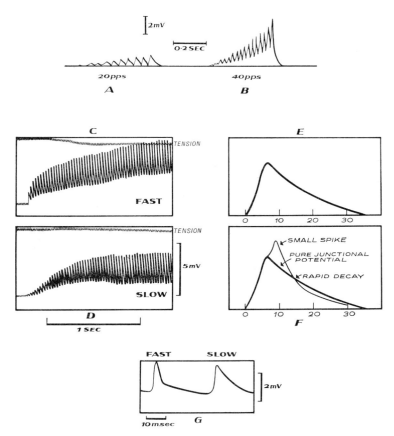

FIG. 40.   Junction potentials.   *A*. Facilitation of junction potentials, *B*. Facilitation and summation of junction potentials at a higher rate of nerve stimulation.   *C*. and *D*. Differences in facilitation, and relationship between depolarization and muscle tension following stimulation of 'slow' and 'fast' nerve axons.   *E*. and *F*. The difference between a pure junction potential and one from which a spike arises.   *G*. Differences in the decay time of junction potentials following ' fast ' and ' slow ' axon stimulation.   (See text for further details.)   (From Hoyle and Wiersma.[261])

Stimulation of the single motor nerve to the dactylopodite abductor of the crayfish *Orconectes* at the rate of 20 pulses per sec. results in a series of discrete junction potentials (j.p's) (fig. 40).[155]

It may be noted that the arrival of the first impulse barely produces any electrical effect but that subsequent j.p's are successively larger and ultimately reach a magnitude of 1–2mV.[155] Such an increase in the response during uniform stimulation is termed *facilitation*. At a rate of stimulation of 40 pulses per sec. the increase in size of successive j.p's is even larger, indicating more facilitation. Moreover, there is no longer a return to the resting potential between pulses, as each new j.p. starts from a little way up the declining phase of the one before it. The increment in height between the peaks of successive j.p's is thus greater than that produced by facilitation alone. The additional increment is termed *summation* (fig. 40B). As a result of the failure of the membrane potential to return to the normal resting level between pulses, there is also a *maintained depolarization*. Its value is the mean height of the j.p's. Occasionally a maintained depolarization can result from long continued stimulation of slow axons when j.p's are not evident (less than 0·1 mV).[261] The height of the j.p's ultimately reaches a maximum level on repeated stimulation.[155] The time taken for this peak size to be reached depends on the rate of stimulation. The higher the stimulus rate, the more quickly the maximum is reached.

As might be expected from what has already been said concerning the differing mechanical responses of different muscles and fibres within muscles to stimulation by the slow and fast axons, considerable variety is also shown in electrical responses. Thus some fibres in the closer of *Panulirus* give very much larger j.p's (up to 10 mV) on slow fibre stimulation at 30 pulses per sec. than do *Orconectes* opener fibres but other fibres from this same muscle in *Panulirus* give no response at all.[261]

An extreme, but rare, situation is exemplified by some of the fibres of the closer of *Pachygrapsus*, where even a single stimulus applied to the slow axon gives a j.p. of 6–10 mV and the height goes up to 15–25 mV on repeated stimulation.[261] This response is more like that to fast axon stimulation in most fibres, but curiously, in these particular fibres there is at most only a very tiny response to the fast axon itself.[261]

THE ELECTRICAL RESPONSES FOLLOWING STIMULATION OF THE FAST AXON

In contrast to slow axon stimulation the first pulse arriving at a fast neuromuscular junction causes a marked j.p. Subsequent stimuli result in facilitation and summation as in the slow axon response, but a higher frequency is usually required to make fast j.p's summate as they tend to have a faster rise and more rapid decay than slow j.p's. In many muscles the fast j.p's are much larger than the slow j.p's of the same fibre. A comparison of slow and fast j.p's is made in fig. 40C and D.

Spike Production

At high rates of stimulation (40 per second and over) a few of the fibres of the closer in *Panulirus* show a modification of the junctional potential response. Much larger potential changes (spikes) are initiated from the peaks of pure junction potentials (fig. 40E, F). Such spikes are distinguished from the junction potentials by their faster rate of decay (fig. 40G). In size the spikes only range up to about 20 mV and so they cannot be considered to be comparable to the all or nothing spikes of crustacean nerve or vertebrate muscle which overshoot the zero potential. Spikes are very rare during slow axon stimulation but sometimes occur when a considerable degree of maintained depolarization has been brought about.[261]

Spike production is most commonly found in those muscles which are capable of fast movement. Thus, in the following series there is both an increased tendency towards spike production on fast axon stimulation and increasing maximum speed of muscle contraction. Extensor of *Panulirus* < closer of *Cambarus* < closer of *Panulirus* < closer of *Pachygrapsus*.

The extensor of *Panulirus* is a muscle designed to display sustained contraction rather than rapid movements. The fast j.p's are only about 1–3mV in height at 10 pulses per sec. but, though they may grow to 15 mV at 50 per sec., they do not show spikes.[261]

The closer of *Cambarus* can react more rapidly than the extensor of *Panulirus* and will give a twitch in response to a single shock.

The electrical response, however, is limited to j.p's which, though they may range to 20 mV, lack proper spikes.[261]

Panulirus is a sluggish animal which does not walk rapidly. Its dactylopodite closer is therefore not called upon to perform rapid movements. At low stimulation rates the fast j.p's are only about 3 mV in size, but at 50 pulses per sec. small spikes are formed.

By contrast, the fast moving crab *Pachygrapsus* has a closer whose response to a single fast axon stimulus is a twitch. The electrical change associated with this consists of large j.p's of 25–50 mV from which rise spikes which may carry the total potential change to 70 mV.[261] The fast axon response fatigues very rapidly in this muscle and indeed it seems to be a fairly general rule that fast axon responses fatigue more quickly than those of the slow axon.

Even within a single muscle the individual fibres may show different speeds of contraction. This is the case in both the flexor[153] and the extensors[17] of *Cancer magister* legs. Some of the fibres respond to stimulation by producing a spike and contracting rapidly; others contract only slowly and do not produce spikes. Morphological differences between the two types of fibre are also apparent. The slowly contracting fibres tend to be smaller than the others and to have their myofibrils clumped together in groups. The larger, phasic, fibres have their myofibrils more evenly distributed.[17, 153] The precise significance of these structural differences is not yet clear, though possibly it is associated with control of the number of active myofibrils. Thus the slowly contracting fibres show summation of contraction and there is some evidence[498] that this effect is brought about by increasing the proportion of active myofibrils. In addition it is probable that differences in the properties of contraction can be correlated with differences in the length of the sarcomeres as in vertebrates. Preliminary studies indicate that fibres in the accessory flexor muscle of the walking leg of decapods which have a distance between successive z-lines of $2-3\mu$ are phasic whilst those which have a sarcomere length of $10-12\mu$ are tonic.[105b]

The spikes discussed above are neither all-or-nothing events nor are they propagated along the muscle membrane. However, propagated spikes are known for a few muscles, as for example the muscle fibres of the muscle receptor organ of the abdominal stretch receptors of crayfish. In vertebrate skeletal muscle, with its single endplate per fibre, such propagation forms an essential part of the spread of excitation. In crustacea it probably has little or no significance because the spike is initiated at so many points along the fibre.[261] The functional importance of a propagated spike is primarily that it initiates a faster and larger contraction than a junctional potential alone.[261] However, if spikes were always produced at the neuromuscular junction, the flexibility of the muscular response would be much reduced.

2. Ion Distribution and Membrane Potentials

There is reason to suppose that the resting potential of crustacean nerves and muscles, and the electrical changes which occur in these excitable tissues on stimulation, are directly dependent on the distribution of the ions across the cell membranes and on the relative permeability of the membrane to different types of ion.

THE RESTING POTENTIAL

The resting potential of crustacean muscles is usually in the range 55–75 mV. Typical muscles, such as the extensor carpopodites of *Portunus depurator* and *Carcinus maenas*, have potentials of about 70 mV when the tissue is bathed by crab ringer solution. If isotonic potassium chloride solution is used instead of the ringer, the result is an instantaneous depolarization to about −1·5 mV.[178] This effect is compatible with the general hypothesis put forward by Hodgkin[253] that the resting potential of the membrane is determined by the equilibrium potential and permeability coefficients of the ions distributed across it. Potassium, being the cation to which the resting membrane is most permeable, has the greatest effect in determining the potential. The permeability to other ions is finite, however, and the resting potential is not usually exactly at the potassium equilibrium potential. The Goldman equation describes the generally accepted relationship between ion distribution, permeability and resting potential:

$$E = \frac{RT}{F} \cdot \log_e \frac{p_{Na}[Na_i] + p_K[K_i] + p_{Cl}[Cl_o]}{p_{Na}[Na_o] + p_K[K_o] + p_{Cl}[Cl_i]},$$

where E is the membrane potential and p is the ion permeability.

Usually the potassium concentration inside cells is high and that in the haemolymph is low. Replacement of the external medium with one containing a high concentration of KCl has the effect of altering the equilibrium potential for potassium across the membrane. Since the membrane potential also is thereby grossly altered, it may be concluded that the muscle is

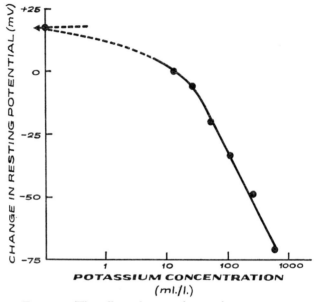

FIG. 41. The effect of external potassium concentration on the resting potential of muscle. *Abscissa*: Potassium concentration in mM/l. *Ordinate*: Difference in resting potential from the normal level present with $K = 12 \cdot 9$ mM/l. (From Fatt and Katz.[178])

usually more permeable to potassium than to the other cations present. The relationship between external potassium concentration and membrane potential is shown in fig. 41.

Replacement of chloride by an organic anion in the medium bathing the claw opener of *Astacus* results in a drop of about 10–20 mV in the resting potential. Chloride may therefore be presumed to play some role in the setting of the membrane level in at least this muscle.

ACTION POTENTIALS

When a squid nerve is stimulated, the potential of the membrane changes for a short time from being negative inside to positive inside. The action potential therefore overshoots the zero potential. It is considered that the production of the action potential is the result of a sudden marked increase in the permeability of the membrane to sodium, so that the potential tends temporarily towards the sodium equilibrium potential. This reverses the normal resting polarity, since sodium has a high concentration outside the axon and a low concentration inside. A similar increase in sodium conductance is thought to be responsible also for the production of the action potential in crab nerve.[253] Replacement of the sodium ion in the external medium by choline has no effect on the resting potential but blocks production of an action potential, and hence conduction of impulses, by the nerve.[178, 255]

By contrast, the potential changes observed during the stimulation of crustacean muscles are less obviously directly interpretable in terms of the sodium conductance. In some cases a potential change following stimulation is very small, and action potentials, even when they do occur naturally, rarely overshoot the zero potential. Furthermore, when sodium in the medium is replaced by the organic cations tetra-ethyl-ammonium (TEA) or choline, the action potential following stimulation, far from being inhibited—as would be the case in nerve preparations—is actually increased in both magnitude and duration.[178] It is probable that the prolongation and tendency towards repetitive firing which occurs in TEA is due to the fact that this substance can substitute for sodium in carrying the inward current whilst at the same time decreasing the membrane conductance, so delaying the recovery repolarization.[547] The divalent ions barium and strontium will also carry the inward current, and, when they are present in the external medium in the absence of sodium, spike action potentials can still be obtained on stimulation of the muscle.[547] TEA, Sr and Ba are, of course, not present in the normal medium; but another divalent ion, calcium, is present, and its role as a possible carrier has been extensively studied.

Calcium does not produce spikes in the normal muscle but, when the internal calcium concentration of barnacle giant muscle fibres is depressed by the injection of EDTA, the muscle membrane behaves during the action potential as though it were a calcium electrode, the size of the spike potential increasing as the external concentration of calcium is raised.[236]   The influx of ^{45}Ca into the fibre during the spike production, after the injection of EDTA to depress the internal calcium level, is more than enough to account in terms of charge for the membrane change.   Thus, for one spike some 35–85 pica mols of Ca, corresponding to 7–17 microcoulombs, enter the fibre.   This is some five to ten times the amount necessary to charge the membrane (whose capacity is only 13–17 $\mu F/cm^2$) by 80 mV.[547]   Indeed the excess is so great that the suspicion must arise that some of this influx represents calcium bound on the membrane and released into the fibre during stimulation.   This is believed to happen in vertebrate tissues, since the rate at which Ca ions enter muscle fibres when they are depolarized by raising the external potassium concentration is not sustained.   However, although Ca, like TEA, Ba and Sr, can undoubtedly carry the depolarizing current under these special conditions, it is doubtful if this is necessarily its normal role in the muscle.   Indeed, since it has been observed that barium-induced spikes are potentiated in the presence of sodium, some authorities consider that, as in other excitable tissues, the entry of sodium is mainly responsible for the depolarization of the membrane.[547]   Such orthodox interpretation of the origins of the junction potentials assumes that the normal graded potentials are of small size, because either there is only a minor increase in the permeability of sodium relative to other ions distributed across the membrane or alternatively the repolarizing outward leakage of potassium on activation occurs too rapidly to allow the membrane potential to shift far in the direction of the sodium equilibrium potential.   The fact that activation of the membrane in the presence of high concentrations of divalent ions results in the production of spike potentials, instead of the normal smaller graded potential changes, is seen as indicating that they decrease the repolarizing leakage of potassium from the fibre at activation rather than that the Ca is the normal carrier.   Support for this idea comes from further observations on one of the pieces of evidence originally taken as suggesting the view that Ca is the normal carrier of the depolarizing currents. This is that the responses of crustacean muscle fibres are greatly reduced or completely abolished on the removal of the external calcium. However, as it is found that in such circumstances the conductance of the fibre membrane is greatly increased so that it becomes depolarized, it appears that this increase in conductance is the reason for the loss of electrical excitability,[547] not the removal of the carrier responsible for inward transmission.

One role of the calcium in the medium, therefore, seems to be to depress the potassium conductance of the membrane. The hydrated ionic radii of TEA, Ba, Sr and Ca are very similar, being respectively 2·89, 2·80, 3·0 and 3·0 Å,[547] making it possible that their effect upon the membrane of producing all-or-nothing spikes, instead of graded potentials, is due to their ability to ' plug ' the pores through which potassium escapes from the fibre without affecting the sodium penetration.[547]   In the absence of sodium they can themselves carry the inward current.

When the calcium concentration of the bathing medium is raised, the potential to which the muscle membrane must be depressed before a spike potential is produced (firing level) moves in the positive direction towards zero.   A larger stimulating current is thus needed to depress the resting potential as far as this and so excite the fibres to produce a spike.[236] Rapid muscle contractions are usually produced in response to spikes in those muscles whose membranes can be excited to all-or-nothing effects.   Hence it is probable that the considerable rise in Ca level in the blood of Crustacea at the time of moult plays a valuable role in depressing muscles and so decreasing the chance that accidental stimulation could cause a rapid contraction capable of distorting the soft new cuticle.

Sodium which enters a nerve or muscle during excitation is removed by a metabolic ' pump '.   In crab nerve an energy-rich phosphatase is considered to be part of this pump, and as sodium is extruded, inorganic phosphate concentration rises.   Between 2·7 and 4·0 Na^+ ions are extruded per energy-rich phosphate bond split.   The mechanism of sodium extrusion is dependent on the presence of Na inside the fibre and K outside.   It is inhibited by ouabain.[20b]

3. Inhibition

Activity in the peripheral inhibitory fibres lessens, or suppresses completely, the contraction of the muscles receiving a supply. Four types of function seem to be served by this unique inhibitory system in the Crustacea:

(1) Common inhibitor axons supplying a number of different muscles probably serve to suppress muscular activity during moult.[554] Otherwise there would be a danger that muscular activity produced at this time by accidental direct mechanical stimulation of the muscle might damage the soft new cuticle. The common inhibitors have not been observed to be active in inter-moult animals.[77]

(2) The specific inhibitor supply to muscles such as the openers and stretcher limb muscles of decapods enables them to show independent activity despite the fact that they share only a single motor axon (fig. 38).[77, 557]

The inhibition of antagonistic muscles can also be accomplished by the peripheral inhibitory axons. For example, when the inside of the claw of *Carcinus* is stroked, the claw is closed and at the same time the opener is reflexly inhibited by its inhibitor axon.[77] In a vertebrate such inhibition of antagonists can occur only centrally, in the central nervous system.

(3) The inhibitory supply to the proprioceptors RM_1 and RM_2 (p. 208) enables the facultative suppression of sensory impulses to the central nervous system when the muscle organ is stretched. This may possibly be of some importance in modifying reflex activity and in providing different types of information from the same sensory systems.

(4) Finally, it has been suggested that the inhibitory supply to the powerful flexor muscles responsible for the tail flick in crayfish may be important in shortening the cycle of flick and extension of the abdomen. The extensor muscles are much weaker than the flexors and hence inhibition of the latter at the end of their effective stroke could bring about a rapid repolarization of the muscle fibre membranes and abolish residual tension, thus assisting the extensors.[289b]

From what is known of the way in which muscles are normally activated, there would seem to be four possible ways in which contraction might be inhibited: (1) by decreasing the permeability of the membrane to ions and so suppressing the electric activity on the arrival of nerve impulses; (2) by decreasing the coupling between membrane changes and contraction; (3) by hyperpolarizing the muscle membrane; (4) by decreasing the membrane resistance so that electrical excitability is prevented. This could be achieved by increasing the permeability to ions such as K and Cl without increasing that to Na or Ca. Any increase in the permeability to Na which might occur on activation could then have little effect on the membrane potential.

Inhibition of muscle contraction can be brought about artificially on crustacean muscle by any of these four methods. Gamma-amino butyric acid and related compounds increase the permeability of the membrane non-specifically to all ions (method 4). High concentrations of calcium in the bathing medium depress the electrical responses of the membrane (method 1). Low concentrations of manganese in the bathing medium decreases the coupling of excitation of the membrane and contraction without affecting the electrical response (method 2).[389] Lowering the potassium concentration in the external medium increases the magnitude of the normal resting potential. A larger change in this potential than normal is then required if contraction is to be initiated (method 3).[389] Thus, as a result of treatment in any of these ways, the stimulating current required to elicit a contraction is increased and the muscle is therefore to be regarded as being inhibited. Stimulation of the inhibitory nerves produces a variety of natural inhibitory responses comparable with some of these artificial effects.

NATURAL INHIBITORY PROCESSES

Three types of peripheral inhibition have been described:

(1) β-inhibition,[77, 262]
(2) α-inhibition,[77, 262]
(3) presynaptic inhibition.

β-inhibition (simple inhibition) is characterized by inhibition of muscular contraction without there being any major effect on the size of the junction potentials. In α-inhibition (supplemented inhibition) both muscular contraction and the excitatory junction potentials produced by motor nerve activity are either decreased in size or suppressed. Both α- and β-inhibition are due to effects occurring at the muscle membrane. Presynaptic inhibition, as the name implies, involves suppression of the release by the motor nerve of the agent which normally excites the muscle membrane at the neuromuscular junction. Presynaptic inhibition occurs only in situations where it may be presumed that the inhibitory nerve endings terminate on, or in close proximity to, the motor nerve terminals, though as yet synapses on the motor nerve have not been observed.

β-inhibition

Suppression of muscular contraction without change in the size of the junction potentials occurs when the common inhibitor to the opener and stretcher of *Cambarus* is active at the same time as the single motor axon.[262]   Inhibition is less complete when fast motor axons are firing at a rapid rate. Thus, in muscles such as the extensor of *Palinurus*, which receive both slow and fast axons but in which the latter do not give rise to spike potentials when they activate the muscle, β-inhibition occurs at low motor axon stimulation rates for both slow and fast axons. At higher rates of fast axon firing, however, inhibition of muscle contraction is incomplete.[262]   Muscle fibres in which the fast axon gives rise to spike potentials, e.g. the claw closer in *Cambarus*, cannot be completely prevented from contracting even at high rates of inhibitor activity.[262]   It seems clear therefore that, though the electrical events are not affected during β-inhibition, their nature does determine whether mechanical inhibition will be complete or not.

α-inhibition

Partial reduction in the size of junction potentials at the same time as reduction in muscle tension has been observed during stimulation of the inhibitor supplying the closer of *Cancer antennarius*,[262] the opener of *Carcinus*[77] and the opener and stretcher of the crayfish.[262, 313]   The fact that the common inhibitor to the opener evokes β-inhibition whilst the specific inhibitor produces α-inhibition makes it possible to compare the mechanical response in both cases.   In most muscles there is little difference,[262] but in the Australian crayfish the muscle tension falls more rapidly at the onset of α- than of β-inhibition.[313]

Attenuation of the junction potentials is maximal only when inhibitory stimuli arrive at the muscle surface shortly (*circa* 1 msec.) before the excitatory impulses in the motor axon.[262]

Of the muscles so far studied, α-inhibition seems to be confined to the opener and stretcher muscles of the Astacura (Nephropsidea) Brachyura and Anomura when stimulated by the specific inhibitor axons.[554]   It should be noted, however, that in practice this may well be the most commonly used form of peripheral inhibition, as these muscles depend for their freedom of independent activity on inhibitor action since they share a single motor axon.

Presynaptic Inhibition

Presynaptic inhibition has been studied in the Crustacea only on the opener of the crayfish.   In this preparation one of the effects of inhibitor action is to decrease the amount of excitatory transmitter material released from the motor nerve axons.[156] As in the case of α-inhibition, for maximal presynaptic inhibition to occur an impulse in the inhibitor axon must arrive near the neuromuscular junction about 1 msec. before the impulse in the motor axon.[156]   Such a time interval would be commensurate with the time required for a chemical inhibitor to diffuse from the inhibitor axon to the motor axon.   The view that inhibition is indeed mediated by such a transmitter is given additional support by the fact that γ-amino butyric acid has an effect which mimics presynaptic inhibition when it is applied to the neuromuscular junction.   Thus in the presence of γ-amino butyric acid the chloride permeability of the presynaptic cell membrane is greatly increased, whilst the quanta of motor transmitter released by the presynaptic cell are decreased.[500b, 500c]

The Mode of Action of Inhibition

Stimulation of inhibitory axons results in the formation of inhibitory junction potentials at the neuromuscular junction. Whereas excitatory junction potentials always tend to depolarize the resting potential of the muscle, the inhibitory potentials may be either depolarizing or polarizing, according to the position of the membrane potential relative to its 'usual' value. The magnitude of the inhibitory potentials is also related to the level of the membrane potential. Both these points are illustrated in fig. 42.

At the start of the experiment illustrated in fig. 42 the membrane potential was 80 mV and the inhibitory potentials were depolarizing. However, as the membrane potential was artificially lowered, the inhibitory potentials became smaller, until at

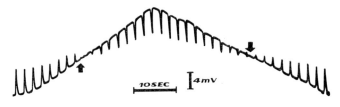

FIG. 42. Trains of inhibitory junction potentials (150/sec. for 0·2 sec.) repeated every two seconds. The resting potential initially at 80 mV was gradually shifted to 61 mV by passing current through an intracellular electrode, and then was allowed to return to the normal level again. The arrows mark the reversal potential at 72 mV. Note the change in polarity of the polarization produced by stimulus at the reversal potential. (From Dudel and Kuffler.[156])

72 mV no potential change occurred on stimulation of the inhibitory axon. Stimulation of the inhibitor following further depression of the membrane potential once more produced inhibitory potentials but of the opposite sign. The effect is completely reversible, raising the potential above 72 mV again being followed by depolarizing inhibitory potentials. A similar effect is found in many other muscles though the precise level at which reversal occurs varies somewhat.[45, 180, 262]

The effect of inhibitory nerve activity is thus to move the membrane potential towards the reversal potential, which usually lies within a few millivolts of the normal resting potential. During inhibitory stimulation the resistance of the membrane is very much reduced; which suggests that the permeability to one or more ions has been increased,[45, 180] so that the membrane potential will tend to move in the direction of their equilibrium potential. The only ions which are present in high enough concentrations and have an ionic distribution across the muscle membrane such that their equilibrium potential would be expected to lie close to that of the normal resting potential, are potassium and chloride. Experiments testing the effect of removal of either potassium or chloride ions from the bathing medium suggest that the inhibitory potentials are largely produced by an increased conductance of the latter ion.[45]   In this respect the muscle junction seems to differ in its mode of inhibition from the sensory nerves of the muscle receptor organs; in these it is an increase in potassium conductance alone that is important.   (Inhibition is further discussed on p. 211).

4. Neuromuscular Transmission

In most neuromuscular systems studied it has been found that the arrival of a nerve impulse at the axon termination is followed by the release of a chemical transmitting agent, which, after diffusing from the axon to the muscle membrane, excites the latter. The general properties of such transmitter processes, based mainly on cholinergic systems, have been summarized by Florey.[192]

(1) The transmitter substance occurs in transit in the whole axon from whose ending it is realeased.

(2) The transmitter is synthesized within the neurone.

(3) Within the neuron the transmitter is stored in an inert form, possibly bound to a protein except when stored in synaptic vesicles.

(4) The transmitter is released from the nerve ending on the arrival of an orthodromic impulse (one travelling in the normal direction along the nerve).

(5) The post-synaptic region possesses an enzyme system able to deactivate the transmitter, but its action in terminating transmitter action may be supplemented by diffusion of the agent away from the site.

(6) The post-synaptic region has receptor molecules which specifically react with the transmitter so as to give an increase in permeability.

Application of these principles to the study of crustacean transmission systems has not reached an advanced state, because comparatively little is yet known of the transmitter substances utilized.

THE TRANSMITTER AT SENSORY NERVE ENDINGS

When acetyl choline (ACh) is applied to the muscle receptor organs of decapods at a concentration of 10^{-6}M, there is an increase in the rate of discharge in the sensory nerve. If the ACh continues to bathe the preparation adaptation occurs, so that the rate of discharge declines exponentially until it reaches a discharge rate somewhat above that of the pre-application level.[191]   It is difficult to conclude that this effect of ACh is specific as it is unlikely that the normal process of activation of the nerve is other than by direct mechanical means.[192] However, ACh has been isolated from crustacean sensory axons[123]; so it is possible that this substance may be utilized at the central synapses of the sensory nerves in the central nervous system.[192]   Compatible with this view is the observation that a substance with properties similar to ACh, and in all probability ACh itself, has been extracted from the central nervous system of a number of Brachyuran and Macruran decapods.[161, 459, 489]   The rate of production of ACh in lobster sensory axons is estimated to be sufficient for 400,000 impulses per hour.[123]

Cholinesterase is also present in considerable amounts in the central nervous system, particularly in the abdominal ganglia.[530]

By contrast with its presence in the sensory axons, ACh is absent from both motor and inhibitory fibres.[192]

THE TRANSMITTER AT MOTOR NERVE ENDINGS

Qualitative as well as quantitative differences between the response of the neuromuscular junctions to slow and fast axon stimulation suggest the possibility that there may be more than one type of motor transmitter. Thus the junction potentials following slow axon activity are longer-lasting, but more readily inhibited, than those produced by fast axon activity.[263, 554]

In vertebrates transmission between axon and muscle is mediated by the release of 'quanta' of acetyl choline, each quantum consisting of many molecules of ACh.[288]   Analysis of the junction potentials of crustacean muscle indicates that they too are produced by the release of quantal packets of a transmitter but, as already mentioned, this is not acetyl choline.   Indeed the nature of the motor transmitter or transmitters is as yet uncertain, though glutamic acid is a possible candidate, since it not only mimics transmitter action but is also small enough to pass through pores in the membrane of the same size as those that allow sodium passage.[193]

Measurement of the membrane potential of the dactylopodite abductor with an intracellular electrode, together with the simultaneous determination of the junction potential at one particular site, indicates that not all the various terminations of a single axon supplying a muscle are releasing transmitter at the same time when the stimulatory frequency is low.[155]   Thus, in one preparation receiving nerve impulses at the rate of one every five seconds, examination of the junctional current flow suggests that the probability of the release of 0, 1, 2 or 3 quanta at a single neuromuscular junction per nerve impulse is in the ratio 270: 115: 55: 20.[154] As the frequency of stimulation is raised, the number of quanta released per impulse is increased, though the size of the individual quantum remains constant.[155]   Release of quanta is thus facilitated as the rate of stimulation is increased.   This facilitation may at least in part explain the facilitation of the junction potentials as the frequency is raised, since each quantum of transmitter produces a junctional potential of about 70 μV.[154]

Fatigue appears to be the opposite of facilitation, a decreasing probability that a quantum will be released at any given axon termination.   This situation could arise if the transmitter present in the presynaptic vesicles is used up faster than it is replaced.

The manner in which facilitation is brought about is still not clear; but it does not appear to represent the invasion of axon terminals which have not previously been reached by nerve impulses.[155]

THE INHIBITORY TRANSMITTER

Considerable attention has been paid to the inhibitory transmitter since it was found that an extract of mammalian brain can mimic its effect.[171, 190, 192]   Fractionation of the extract indicates that the effect is due largely to the substance gamma-amino butyric acid (GABA).

$$\text{GABA} \quad \begin{array}{c} \text{H} \\ \text{H} \end{array}\!\!\!> \text{N} - \overset{\displaystyle \overset{\text{H}}{|}}{\underset{\displaystyle \underset{\text{H}}{|}}{\text{C}}} - \overset{\displaystyle \overset{\text{H}}{|}}{\underset{\displaystyle \underset{\text{H}}{|}}{\text{C}}} - \overset{\displaystyle \overset{\text{H}}{|}}{\underset{\displaystyle \underset{\text{H}}{|}}{\text{C}}} - \text{C} \!\!<^{\displaystyle \text{O}}_{\displaystyle \text{OH}}$$

Treatment of the muscle receptor organ (*q.v.*) of the crayfish with this substance, like stimulation of the inhibitory nerves, produces little change in the membrane potential of the resting sensory axon but returns that of a previously active axon to the reversal potential level.[311]   GABA appears to act by producing a selective increase in the permeability of the nerve membrane to potassium and chloride ions[168, 311] thus clamping the membrane potential at the reversal level.   A number of ω-amino-acids and ω-guanido acids also produce a comparable, though smaller, effect.

The action of GABA is localized to the region of the dendrites and sensory cell body, i.e. the region where the inhibitory axon terminates.   A concentration of GABA 1000 times that which blocks sensory nerve activity when applied at the cell body has no effect on transmission if applied along the axon.[311]   The action of GABA at the proximal end of the axon is not, however, necessarily limited to any specific inhibitory sub-synaptic site as it has also been found to cause inhibition of the lateral stretch receptor of the seventh thoracic segment of the lobster,[311] an organ which lacks the inhibitor nerve.[3]   By inference, therefore, it must be assumed that the whole surface of the dendritic and soma region is more sensitive to the presence of the inhibitor than is the axon.

There is some evidence to suggest that the natural inhibitor transmitter and GABA are chemically related.   Thus, when a stretch receptor preparation is bathed in a medium to which GABA has been added, the degree of blockage of transmission by the sensory nerve gradually declines if the medium is not stirred, but returns to the initial level on stirring.[168, 311]   This result could be explained if the GABA were being neutralized at the cell surface by the system normally responsible for deactivation of the inhibitory transmitter.   This would be likely to occur only if the two substances were related in their chemical properties. Further evidence of this relationship is provided by the fact that both natural inhibition and Factor 1 (an extract which contains GABA) are deactivated by picrotoxin. It may be recorded in passing that ethyl alcohol also partially blocks Factor 1, a finding which gives sophisticated support to the popular conception that alcohol ' removes inhibitions '.[171]

These various pieces of evidence all point to a similarity of action between GABA and the natural inhibitory transmitter, but despite both this and the fact that GABA has been extracted from crustacean nervous systems,[157, 305, 311] it is possible that it is not the natural transmitter. A considerably more potent material (Substance 1), also blocked by picrotoxin,[194] has been extracted from the peripheral axons of crabs.[192, 194] This behaves like the natural inhibitor when assayed on the heart, dactylopodite adductor and stretch receptor organ but in terms of the magnitude of the effect produced is very much more effective than GABA. Thus the inhibitory nerves to the heart of *Homarus* are estimated to contain the equivalent of the activity given by 300,000 μgm GABA/gm wet weight of tissue—an obviously impossibly high level if the substance were actually GABA.[192] The true nature of Substance 1 remains to be elucidated, but it seems to be ninhydrin positive.[194] The suggestion that it may be aspartic acid is compatible with this observation.[193] Despite the discovery of Substance 1, it seems clear that GABA must play some part in the natural inhibitory process, if only for the reason that, whereas it is present in inhibitory fibres in the lobster leg, it cannot be detected in the motor fibres.[305] Whether its role is that of precursor or part of the natural transmitter remains, however, to be established.

5. The Coupling of Electrical Changes of the Muscle membrane and the Contraction Process

Several theories have been put forward in attempts to explain the nature of the stages involved in the excitation of the contractile processes following the arrival of a nerve impulse at the neuromuscular junction. Among these the most important are:

1. The potential change initiated at the neuromuscular junction causes an internal current flow between points at different

potential levels, and this current excites the contractile elements.

2. The contractile mechanism is directly linked to the membrane potential, so that contraction is initiated when the membrane has been depolarized to a threshold level.[180]   The depolarization is presumed to be conveyed directly to the contractile elements by intuckings of the membrane.[272]

3. Depolarization of the muscle membrane beyond a given level results in the release of a transmitter substance which diffuses to the contractile elements and excites them.

Before attempting to assess which of these theories most closely approximates to the true situation it is necessary to outline some of the properties of the coupling process which have been experimentally established.

(a) In many muscles it is found that contraction is initiated after a certain degree of depolarization has occurred, and that further depolarization is associated with a rise in muscle tension (fig. 43).[180, 263, 390, 498]

(b) If the membrane of the crayfish contractor epimeralis is kept depolarized just beyond the threshold level, the tension developed by the muscle continues to rise for as long as 20 seconds.[390]   Any changes made in the membrane potential during this time are however immediately reflected by a change in tension, transient decreases in depolarization producing a temporary decrease in tension[390] and transient increases a temporary rise.   Thus it would seem certain either that the level of membrane potential is itself of major importance in initiating processes leading to contraction or that the current flow across the membrane is carried by an ion which excites contraction.

(c) Direct depolarization of the muscle membrane with external micro-electrodes results in a contracture of the fibre when the electrode tip is placed in the region of the A band close to the boundary with the I band but has no effect in other regions.   The contracture is local and takes the form of the shortening of one I band (presumably by the sliding of the filaments into the stimulated A band) so that the Z line comes to lie adjacent to the A band.   The neighbouring I band is not affected (fig. 44).[272]

This type of contracture is similar to that obtained in lizard muscle fibres but differs from that in the frog where contracture

occurs only if stimulation is at the Z line.[272]   At the level of the
A-I band junction in the crayfish and lizard,[547, 272] and the Z
band in the frog,[417] there are transverse vesicles which it would be
plausible to suppose from their position, in relation to the position
of the electrode necessary to cause contracture, may be concerned
in the conduction of excitation to the contractile elements.[272]

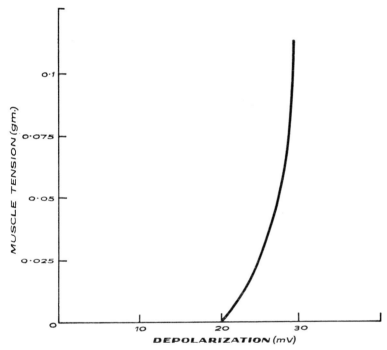

FIG. 43.   Increasing muscle tension with increased depolarization
above the threshold level.   Resting potential before depolarization
75 mV.   (Based on a figure by Orkand.[390])

(d) Any possibility that contraction of crayfish muscle is caused
directly and solely by a spread of electrical activity during membrane depolarization is difficult to equate with the behaviour of the
closer muscles of the brachyuran *Randallia*, Anomuran, *Blephari-
poda*[263, 559] and of fatigued *Cambarus* closer.[263]

On stimulation, these muscles respond in a normal manner at
first.   In *Randallia* closer, for example, the junction potentials

and depolarization produced by stimulation of the fast axon are greater than those produced by stimulating the slow axon, and the contraction produced by 10 impulses per sec. is greater when the fast axon is stimulated.   After a little time stimulation of the fast axon ceases to evoke any contraction, even though the junction potentials and maintained depolarization of the muscle

A   I   Z   I   A   I   Z   I   A

BEFORE STIMULATION

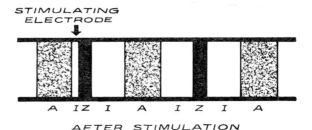

STIMULATING
ELECTRODE

A   IZ   I   A   I   Z   I   A

AFTER STIMULATION

FIG. 44.   The local contracture of muscle fibres after stimulation at the junction of I and A bands.   (Drawn from a text description in Huxley and Taylor.[272])

membrane are developed as before.   Stimulation of the slow axon at this time produces contraction.   This is the well known ‘ paradox ’ found by Van Harrevald and Wiersma (1935).   Recent re-examination of *Randallia* muscle has shown the presence of thin muscle fibres at both ends.   These fibres, which give a large contraction on slow axon stimulus but little response to the fast axon, are responsible for the paradox effect, since they can still contract when the other fibres of the muscle have fatigued.[17b] The ‘ paradox ’ is therefore no longer a paradox.   The point that requires stressing, however, is that during the process of fatigue the electrical excitation of the membrane of the main muscle fibres may remain normal though the contraction process is exhausted. Membrane depolarization cannot therefore be directly responsible for exciting the contractile elements.

(e) Additional support for this conclusion is provided by the response of muscle fibres to the presence of manganese. Low concentrations of this ion do not affect the electrical response of the muscle to nerve stimulation but do block contraction. Nevertheless the muscle returns to normal so rapidly on removal of the manganese that it seems improbable that the effect can be due to penetration within the fibre and direct inhibition of the contractile elements.[389] It is more likely that the ion interferes with the release of some activating material from the membrane itself. If these interpretations are correct, the membrane potential may play a role in the initiation of contraction by bringing about the release of some intramuscular transmitter substance from specific sites in the membrane.

(f) The calcium ion is also implicated in the excitation of muscle contraction and may be this transmitter. Vertebrate muscle fibres placed in calcium-rich medium show contracture at any damaged points where the protoplasm comes into contact with the medium.[243] Also, injection of calcium—but not of potassium, sodium, or magnesium—causes marked contraction of muscle fibres,[244] and it has been observed that there is a marked increase in the rate at which calcium ions enter muscle fibres while they are being stimulated.[34, 236] Furthermore, as the rate of calcium entry during depolarization of the muscle membrane, produced by raising the external potassium concentration, is not sustained, it is probable that much of the calcium penetrating the depths of the fibres comes from bound sites on the muscle membrane.[34] Comparable experiments on all these aspects have not yet been performed on crustacean muscles, though it has been found that, as in the vertebrates,[80] micro-injection of calcium causes contracture of the fibres of *Maia*.[71]

The effects of injection of calcium have provided additional evidence for suggestions that the natural Ca^{++} concentration in the sarcoplasm of resting *Maia* muscle is of the order of 0·01 – 0·1 μM, and that on excitation the level rises to 0·1 – 1·0 μM, later falling back, on relaxation, to the resting concentration.[716]

Such a rise in concentration might be responsible for initiation of contraction.

These calcium levels are close to those required respectively to depress and excite the interaction of Actin and myosin, and A.T.P.ase activity. Isolated Actomysin systems from *Maia squinado* are dissociated if the calcium level falls below 10^{-7} M, whilst both mechanical and A.T.P.ase activity reach a maximum at 10^{-5} M. From the points raised in $(a)-(f)$ above it appears that the third of the three hypotheses is nearest to the truth, though this will not entirely account for the observed phenomena.[244b]

The recent discovery that in crustacean muscle there is a system of transverse tubules lying on either side of the Z line in the region of the junction of the A to I bands and that the wall of this tubular system is selectively permeable to Cl ions [220] does, however, make possible a hypothesis of excitation-contraction coupling incorporating elements of the three theories.

Electron micrographs indicate that the walls of the transverse tubules are in connection with the muscle fibre membrane via a system of radial tubes,[215] and that the lumen is in connection with the outside medium. There is thus a spatial separation of the regions of the cell membranes with different specific ion permeabilities. The cell membrane acts as a K or Na battery according to whether it is in the resting or active state, whilst the tubular membrane can act as a Cl battery. Thus, when the membrane is depolarized on activation, the entry of Cl ions from the tubule lumen into the cell proper could complete the circuit resulting from the entry of positively charged ions (Na) at the cell surface. Since only chloride ions can traverse the tubular membrane, it is likely that positively charged ions conducting the current intracellularly, particularly such divalent ions as Ca^{++}, will accumulate in the region of the transverse tubular membrane. This concentration at the junction of the A to I bands (which have already been mentioned as being an area particularly sensitively involved in the contraction mechanism) could be responsible for triggering the contraction and breakdown of A.T.P.[215] If calcium is released from the cell membrane and diffuses inward when the latter is activated, and if the amount of calcium released is related to the degree of depolarization, this would offer an explanation both of the relationship between membrane potential change and degree of muscle contraction and also of the observation [498] that there is a tendency for the more peripheral fibrules to contract before those farther from the point of muscle activation.

6. The Central Nervous System and Segmental Nerves

In its most simple expression the crustacean central nervous system consists of a dorsal brain representing several fused ganglia

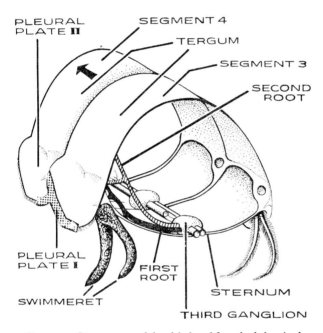

FIG. 45.   Stereogram of the third and fourth abdominal segments of a crayfish, showing the sensory innervation of the right half of the third ganglion.   *Dark shading*: First root's field; *Light shading*: Second root's field. Arrow shows the position of the abdominal stretch receptors for the joint between the fifth and fourth segments.   (From Hughes and Wiersma.[268])

and paired circum-oesophageal commissures connecting the brain to widely separated paired ventral longitudinal cords which have ganglia and cross-binding commissures in each segment.   The whole arrangement is thus ladder-like. In higher forms the nervous system is more compressed, with fusion of ganglia occurring in association with fusion of segments so that in an extreme case, such as that found in Brachyura, all the ganglia of the ventral cords are fused into a single mass. Even in the last condition, however, the segmental arrangement is still recognizable by the nerve roots leaving the ganglionic mass (fig. 46). Typically, as in the abdominal ganglia of *Astacus*, three pairs of nerve roots run from each ganglion. The first two pairs in each segment are mixed nerves (containing both motor and sensory elements) the third root carries only motor fibres (fig. 45).[553]

FIELDS OF INFLUENCE OF THE SEGMENTAL NERVES

Knowledge of the spheres of influence of the nerves running in the three segmental roots is extremely limited, even semi-detailed analysis being confined to the crayfish. The sensory fields covered by the first segmental nerve include the pleural plates, swimmerets and ventral surface of the same segment. Touching the hairs in any of these regions, or moving the joints of the swimmerets, results in sensory activity in the nerve.

The second segmental nerve supplies mainly the segment posterior to the one in which its ganglion lies. Sensory moities include the tactile sense of the tergum and proprioception from the abdominal stretch receptors, RM1 and RM2. Tactile sense in the latero-posterior border of the preceding segment's tergum is also mediated by this nerve (fig. 45).

Motor fields

The motor nerves of the first abdominal root supply the muscles of the segmental appendages, those of the second root innervate the majority of the intersegmental muscles[554] with the exception of the abdominal flexor muscles, which are supplied via the third root.[282] The nerves travelling by the second root to the dorsal muscles supply the muscle of the next most posterior segment,[237] thus showing a displacement comparable with that of the sensory nerves in the same root.

Inhibitory nerves

It is probable that inhibitory nerves run in the same nerve root as the motor fibres to a muscle, though physiological evidence on this point is lacking.

FIG. 46. The thoracic ganglionic mass of *Carcinus*, showing a single nerve cell which sends axons to four limbs. (Modified from Bethe.[33])

Histological preparations show that one efferent cell on the visceral nerve mass of *Carcinus* sends axons down the nerve supply to several successive legs.[33] This may perhaps be the common inhibitor to the leg muscles (fig. 46).[554]

CONNECTIONS WITHIN THE CENTRAL NERVOUS SYSTEM

The most simple nervous circuit possible is that of the reflex arc where a sensory axon makes a direct synapse with a motor cell. A probable example of such a simple arc has already been described in discussing the response of the nerves and muscles of the limb to passive movements of the joint between the propodite and dactylopodite. Movement of the joint in the direction normally produced by contraction of the opener muscle results in reflex activity in the slow motor axon to the closer muscle. This reflex arc cannot, however, involve only a connection between sensory

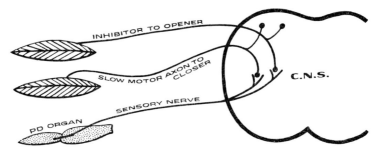

FIG. 47. A reflex arc.

terminations and motor cell, as activity in the specific inhibitor to the opener is itself inhibited at the same time.[78]   Fig. 47 illustrates a possible connection within the central nervous system for such a reflex.

Some primary sensory axons may branch in the central nervous system and send ascending, descending, or both ascending and descending, fibres along the nerve cord; but it is most likely that these ultimately terminate in synapses with interneurones (cells restricted to the central nervous system), rather than on motor fibres with cell bodies in neighbouring ganglia. The very limited number of cells in the crustacean central nervous system (only some 12,000 axons in the abdominal connectives and 2,000 in the second oesophageal commissure[268]) make it necessary that most interneurones should make a number of synaptic connections. The ability of the central nervous system to analyse and integrate the sensory information is obviously dependent on the nature of its connections to the sensory supply. Various types of interneurone have been located[556] illustrating different levels of integrative function:

1. Interneurones which respond only to localized sensory activity in a single segment e.g. two which fire when the fourth and fifth swimmerets are in a backward position.

2. Interneurones active after stimulus of similar sense organs in several regions served by the axons passing in a single nerve root, e.g. interneurones responding to touch of the dorsal region of one segment and the second pleural plate in the segment anterior to it.

3. Interneurones integrating more than one type of sensory input within a single segment, e.g. interneurones responding to touching the hairs on the pleural plate and also to movement of the swimmeret joints.

4. Interneurones integrating the sensory input of more than one segment.

Less specific information is conveyed by interneurones responding to sensory input from several successive segments. Thus interneurones are known which are active following the touching of the dorsal hairs of two segments, whilst others respond to three segments. Some measure of localization is potentially possible even where there is input from several segments if the fields of different interneurones interact, provided that the different combinations of active interneurones can be assessed (fig. 48).

There are three possible ways (fig. 49) in which sensory nerves entering the roots of different ganglia could make connections with a single interneurone: (*a*) the primary sensory nerves themselves might run along the central nervous system to the interneurones; (*b*) the sensory nerves might make separate synapses with interneurones which can then have synapses with the common interneurone; (*c*) an interneurone might make synapses in successive ganglia with incoming sensory axons.

An example of the first situation is found in the case of the sensory inflow from the slowly adapting RM1 stretch receptors. The primary sensory axons enter the central nervous system and then, after branching in the ganglion at which they first enter, descend along the cord to synapse with the common interneurone in the sixth ganglion (fig. 50).[268]

A number of other primary fibres, particularly those from the touch receptors of the pleural plates and the proprioceptors of the swimmerets, have also been located in the connectives posterior

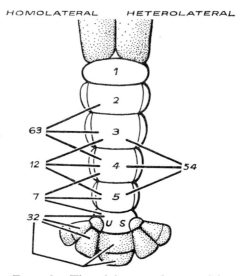

FIG. 48. The abdomen of a crayfish, showing the areas innervated by inter-neurones responding to touching the dorsal hairs of three adjacent abdominal segments. The numbers indicate the reference numbers of particular axons in the central nervous system. (From Wiersma and Hughes.[556])

to the ganglion in which they enter,[268] suggesting that this type of connection is not uncommon. The form of integration achieved will depend on the properties of the synapses, the threshold levels, the excitability, etc.

Type B connections have not been positively demonstrated, but type C connections are also indicated by the discovery of interneurones which respond to stimulation of particular sensory areas (e.g. the hairs on the pleural plate) in several neighbouring segments. Impulses travel in both directions along such inter-neurones, anteriorly if the posterior segments are stimulated, and posteriorly if the anterior segments are stimulated.[268] A feature of such interneurones is that simultaneous stimulation of anterior and posterior segments results in impulses travelling in opposite directions along the axon and cancelling each other out when they

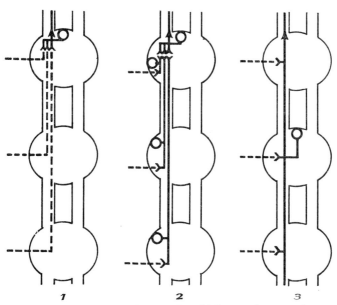

FIG. 49. Three possible ways in which neural connections might be made which would result in a single interneurone firing when sensory axons in three separate segements are stimulated. - - - - Primary sensory axons, ——— Inter-neurones. (From Hughes and Wiersma.[267])

meet. Loss of sensory information as a result of this is unlikely; for not only will there be synapses between the first interneurone and others of the neuropile at several different points, but also the interneurones associating information from several different segments are supported by the interneurones more specifically concerned with single segments.

In the crayfish, physiological evidence suggests that at least some of the type C interneurones receiving sensory axons from several segments have synaptic connections in one or more ganglia with their homologue on the other side, ensuring bilateral distribution of sensory information. Where such synapses are made in several ganglia, there can be no blockage of end-to-end transmission, even if one of the axons of the interneurone is severed between

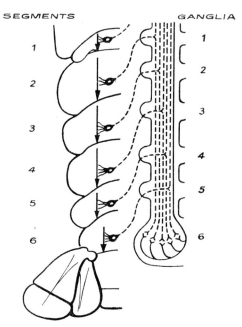

FIG. 50. Diagram of the path of sensory fibres from the homolateral slowly-adapting muscle receptor organ in the abdomen. One half of the cord, from the first to the sixth ganglion, is shown. - - - - Primary sensory fibres, ———— The interneurone of the sixth segment. Arrows indicate the joint which, when flexed, excites a given receptor. (From Hughes and Wiersma.[268])

two ganglia, since the damaged area can then be by-passed by impulses travelling in the heterolateral interneurone.

The presence of some interneurones having only one or a limited number of synapses with their opposite number is indicated by observations that sometimes when a single connective is cut sensory stimuli from the opposite side of the body still produce impulses in the interneurone anterior to the cut, but not posterior to it.

In addition to variations in the number and type of connections made to the sensory primary axons, the interneurones show a considerable range of responses to stimulation. The threshold of stimulus required to excite different interneurones varies, as does the length of time the interneurone remains active during stimulation. Some are readily excited, and remain active as long as the sensory axons supplying them are firing. At the other end of the scale Wiersma and Hughes[556] have found an axon whose threshold is so high and which fatigues so rapidly on stimulation that it is rarely active for long. They have nick-named it the ' quack ' fibre.

Interneurones responding to either tonic (long-term) or phasic (transitory) stimulation of sense organs provide further means of analysing sensory information.

In the vertebrate central nervous system most synapses are axo-somatic (i.e. between the axon of one cell and the cell body of another) or axo-dendritic (between the axon of one cell and dendrites of another). By contrast, many of the synapses in decapods are between axons.[289] Any one interneurone can synapse with number of other axons. The potentials produced at a synapse by a stimulation of the pre-synaptic cell are graded in size and, if small, will not be sufficient to promote a propagated potential in the post-synaptic fibre. However, it is found that if such sub-threshold synaptic potentials are produced at several of the synapses of the post-synaptic cell then they may sum together and initiate a spike discharge. Such a system can clearly play an important role in integrative functions within the central nervous system.

In at least some interneurones the production of the spike discharge does not destroy the synaptic potentials which brought it about. The latter can thus initiate further spikes. In these cases, therefore, the number of spikes in the post-synaptic axon can exceed that in the pre-synaptic cell.

Another feature of importance in integration is that activity in some pre-synaptic fibres can inhibit spike discharge in the post-synaptic axon produced by activity of other pre-synaptic fibres.[421] Given these various properties, the individual neurone has considerable potential for integration of the input to it.

14

Summarizing the main points, multiple stimulation by one or more presynaptic axons may be necessary to excite one branch of an interneurone. Even when this branch is active, there may be no spread of activity to the main axon unless other branches are also active. Additional control over the response of the main axon may be exercised by presynaptic inhibitory endings. Therefore, even if the main axon has only a single type of response when active, the types of stimulus required to evoke this response are potentially very varied. Clearly the somewhat stereotyped behaviour of Crustacea is due, not to any lack of complexity of means of integration in the individual cells of the CNS handling the sensory input, but rather to their limited numbers.

So far we have discussed only the integration of the sensory input to the central nervous system. Some of the neurones responsible for co-ordinating the motor output have also been located.

Central Integration of Motor Output

Co-ordination of the motor system as a whole demands that when a muscle is to be excited its motor neurones be active whilst those of antagonistic muscles be suppressed; its own inhibitor axon must be quiescent and those of antagonistic muscles active. An integrated output showing these features can be initiated in some cases by stimulating a single interneurone (command fibre) in the abdominal nerve cord. An example of the operation of such a command fibre has been demonstrated on the control of the postural muscles of the crayfish abdomen. The slow extensor and flexor muscle on one side of an abdominal segment are each innervated by five motor and one inhibitory axons. Stimulation of the appropriate command interneurone for flexion or extension brings about the following responses. Activity in the flexion command axon produces excitation of the five flexor motor neurones, inhibition of the flexor peripheral inhibitor neurone, excitation of the extensor peripheral inhibitor and central inhibition of the five extensor motor neurones. Precisely the opposite effects are produced by stimulation of the extensor command interneurone. The activity of a single command interneurone is capable of organizing in this way the output of over 100 separate efferent axons.[289d]

Several interneurones may produce the same type of extensor or flexion command response to a single segment, but they differ in the degree to which the area of their influence is spread to neighbouring segments of the abdomen. Control over the activity of these various command fibres will thus determine the posture of the abdomen as a whole.

SPONTANEOUSLY ACTIVE NEURONES

Intracellular recordings made in the region of the sixth abdominal ganglion indicate that there are some interneurones there which are apparently discharging spontaneously at rates of one pulse every 10–30 msec., or in some cases even more frequently. These cells are probably pacemakers, and the evidence at present available suggests that their rhythmic behaviour is endogenous rather than being brought about by external stimulation. The rhythm can, however, be reset by presynaptic impulses in some cases; and this suggests that the output can be modified by other cells.[421b]

GIANT FIBRES

Forms such as the Natantia and Astacura, whose principal escape reaction involves a rapid flip of the tail, have the action mediated by giant axons in the central nervous system. Two pairs of giant axons run unbranched along almost the whole length of the central nervous system. The medial pair are uninterrupted but the lateral pair show segmentally arranged septa at least in the abdominal region, indicating their multicellular origin.[282] The median giants make synapses in the brain with sensory cells in the anterior part of the body, the lateral ones with posterior sense cells. Fairly severe sensory stimulation is necessary to initiate activity of the giant axons. Once initiated, however, activity passes rapidly along the fibre almost unimpeded, the septa of the lateral giants imposing a delay of only 0·1–0·2 milliseconds, as opposed to the synaptic delay of about 1 millisecond at a neuromuscular junction.[534] Lateral commissures bridge the lateral giants segmentally. These also appear to have septal junctions separating the two laterals, but again the transmission lag at this junction is extremely short.[534] As a consequence of the presence of these commissures, any region of one lateral giant which is temporarily non-conducting for any reason can be by-passed by the impulses. The safety factor for end-to-end transmission is thus much enhanced. In decapods the giant axons make synaptic connections with the seven pairs of motor axons supplying the flexor muscles of the abdomen (one pair in each of the first five abdominal ganglia and two in the last). The cell body of each motor axon lies in a ganglion, and its axon crosses

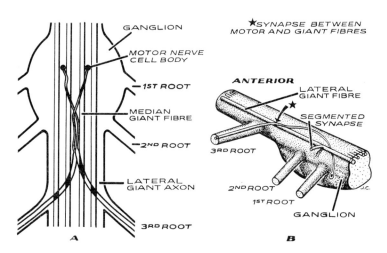

FIG. 51. Ganglion and short section of ventral nerve cord, showing the relationship between the four giant axons and motor nerve axons. ■ Areas where there are synapses between axons. (A. Modified from Johnson,[282] B. Modified from Furshpan and Potter.[213])

laterally to make a synapse with both the heterolateral giant and the medial giant before leaving the central nervous system to supply the muscles (fig. 51).[287] It is apparent, therefore, that these motor axons can be excited in at least three ways: (1) activation by their own cell body, (2) activation via the median giant axon, and (3) activation via the lateral giant axon.

When the giant axons are activated, all the abdominal flexor muscles will contract simultaneously and the animal will display the escape reaction. Independent action of the muscles is achieved by stimulation via the motor body.

Impulses pass the septa of the lateral giants with equal facility in either direction, and the same is true of most of the commissural septa between the lateral giants.[534] It is obviously important, however, that transmission should be one way only at the synapses between the giant axons and the motor axons; otherwise, any activity in the latter might excite the former and hence produce an undesirable escape reaction.

Like the transmission at the septa that between the giant fibres and motor axons is electrogenic and has an extremely short delay. Unlike the septa however, the lateral synapse behaves as if it were a rectifier allowing the passage of positive current only in the direction of a giant axon to motor axon.[213] This serves to prevent excitation spreading from the motor axon into the giant axon.

It has been customary in the past to regard the vertebrates as having the fastest rate of transmission of impulses along nerves. However, the giant axons of the prawn *Penaeus japonicus*, have now been found to transmit impulses at up to 210 metres a second, a rate exceeding even that of the cat sciatic nerve.[516b] Speed of transmission is associated with a high resistance and low capacitance of the nerve membrane, together with saltation of the nerve impulse between nodes along the nerve. The membrane resistance in the prawn giant axons is $3 \times 10^4 \ \Omega/\text{cm}^2$, as compared with a vertebrate maximum of about $1 \cdot 6 \times 10^5 \ \Omega/\text{cm}^2$; and the capacitance of the prawn membrane is $1 \cdot 35 \times 10^{-8} \ \text{F/cm}^2$, the vertebrate maximum being about $5 \times 10^{-9} \ \text{F/cm}^2$. Morphological structures resembling the nodes of Ranvier of vertebrates cannot be seen in the prawn, but the electrical evidence suggests that some regions with comparable physiological properties, permitting saltation of the impulse, must exist.[316b]

CENTRAL INHIBITION

As in vertebrates, interneurones with an inhibiting function seem to be present in the central nervous system of Crustacea, though much of the evidence is indirect as they have as yet not been intensively studied. It has already been mentioned that inhibition of the opener muscle during passive opening of the claw is achieved almost solely by peripheral inhibition of the muscle, with little change in the activity of the motor axon.   This form of inhibition is appropriate only to muscles, such as the opener and stretcher, which have specific inhibitory fibres.   Reflex

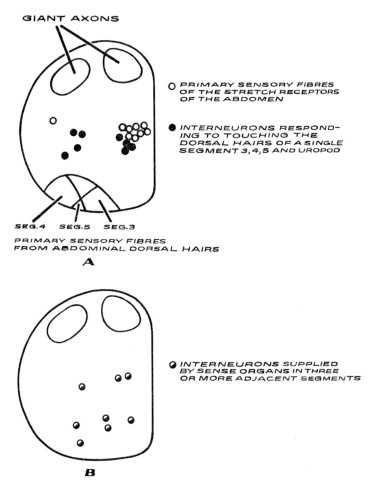

FIG. 52. Transverse section of crayfish nerve cord (unilateral), showing the distribution of nerve axons of various types. *A*. Bundles of primary sensory fibres and interneurones responding to touch of a single segment. *B*. Scattered arrangement of interneurones innervated from more than three segments. (Modified from Wiersma and Hughes.[556])

inhibition of the other leg muscles, which only share an inhibitor, must be achieved by inhibition of the motor axon.[77] Some central control of the inhibitory process can be exerted even in the case of the opener muscle; for it is found that, should there be a burst of discharges in the motor neurone, the activity in the inhibitor declines for a short time. This implies that the inhibitor can itself be inhibited centrally.[77] Central inhibition of transmission between the giant axon and the motor axons supplying the flexor muscles of the abdomen is also known to occur.[214] A particularly interesting feature of this inhibition is that it appears to be mediated by a transmitter substance whereas excitation is electrogenic.[214] Inhibition of interneurones is also possible. An interneurone has been located in the abdominal region which is active when the fifth swimmeret is in a backward position. This activity is inhibited if the dorsal hairs of the same segment are touched.[556]

DISTRIBUTION OF AXONS WITHIN THE
CENTRAL NERVOUS SYSTEM

Well marked tracts composed of axons of similar function are less obvious in the crustacean nerve cord than in the central nervous system of vertebrates, if the situation found in the crayfish can be taken as being typical of the group in general. In the crayfish the only axons grouped into tracts are those of primary sensory fibres which run along the cord and of those interneurones which supply only a single segment (fig. 52). Interneurones being supplied by sensory axons from more than one segment tend to be rather widely scattered in the connectives, though any particular fibre is always located in the same place in different individuals. The presence of 'silent' zones around the giant axons in isolated abdominal preparations suggests the possibility that the descending fibres from the brain may be grouped in this region.[556]

8 : Sense Organs

Sensory receptors fall into two classes, endoreceptors and exoreceptors. The former respond to stimuli arising within the body, the latter to features of the environment or effects occurring at the body surface.

1. Endoreceptors

Endoreceptors are responsible for notifying the central nervous system of the state of various systems within the body and of any changes which occur therein. We may expect that systems are present which respond to changes in body volume, chemical and physical composition; but these are as yet unknown. Indeed the only endoreceptor system so far studied in Crustacea is the proprioceptive system, which is responsible for signalling changes in body or limb posture.

PROPRIOCEPTORS

The maintenance of equilibrium and the capacity to walk or swim necessitate that the central nervous system be constantly informed as to the relative positions of different parts of the exoskeleton, the tension in muscles and the rate of movement of movable parts. These requirements are subserved by the proprioceptive organs. Three types of proprioceptive organ are known in the Crustacea:

1. Muscle receptor organs
2. ' N ' cells
3. Connective tissue stretch receptors—chordotonal organs and non-chordotonal organs.

A fourth type, functionally associated with the central nervous system, is suspected on physiological grounds, but has not yet been physically located.

1. *The Muscle Receptor Organs*

Proprioceptive organs comparable in structure with the intra-fusal muscle proprioceptors of vertebrates are not found in crustacean muscles but a limited number of muscle receptor organs of an essentially similar type are situated in the dorsal part of the last two thoracic segments and in all the abdominal segments of a number of Decapoda.[2, 3, 4, 5, 6, 195]   As these organs are also present in Anomura, Stomatopoda,[416] in the isopod *Ligia*,[6] and in the mysid *Praunus*,[6] it may be supposed that they are of general occurrence in the Malacostraca.   As might be expected, they do not occur in the reduced abdomens of brachyuran forms.[416]

In the abdomen of *Homarus* and *Palinurus* the muscle receptor organs are paired structures lying at the base of the superficial dorsal muscles somewhat lateral to the mid-line (fig. 53*A*).   They are, however, quite separate from the effector musculature.   The muscle of each receptor organ unit has its origin on the integument at the leading edge of one segment, and is inserted on the integument near the anterior end of the preceding segment (fig. 53*B*). The arrangement is such that the muscle will be stretched when the abdomen is flexed (fig. 53*C*), and so the organ is potentially able to provide information concerning the degree or rate of flexion.   The two muscle units of each pair differ somewhat in their length and histological appearance, the more lateral of the two (RM_1) being shorter and slightly thinner than the other (RM_2) and with a less undulating arrangement of its myofibrils.[2, 195] Both RM_1 and RM_2 usually have a single sensory nerve cell associated with them[2, 195] though accessory cells, which also are apparently sensory, are present in *Leander*.[6]   The cell body of the sense cell lies close to the muscle, and from it a massive arborizing system of dendrites penetrates the muscle tissue.   In the terminal region, the dendrites supplying RM_1 run parallel to the muscle fibres for a considerable distance. RM_2 has no such definite arrangement of the dendrite endings.[195]

The details of the nerve supply to the organs differ from one species to another.   Each abdominal RM is supplied by one large motor axon and by the branches of two accessory (inhibitor) fibres in *Homarus*,[2] but in *Cambarus* only RM_2 receives a large motor fibre, RM_1 in this case being supplied by a bundle of fine

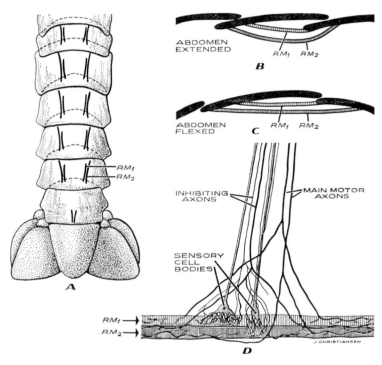

FIG. 53. The abdominal muscle receptor of the lobster. *A.* Dorsal view of the abdomen, showing the paired muscle organs RM_1 and RM_2 in each segment. *B.* and *C.* Vertical longitudinal section of the dorsal body wall, showing how the receptor organs are stretched when the abdomen is flexed. *D.* Detailed structure of RM_1 and RM_2 showing the innervation. (Redrawn from Alexandrowicz.[2])

efferent fibres.[195]   The accessory fibres in *Homarus* are of two types; thin ones supply only the region where the dendrites of the sensory cells terminate, whilst thick ones supply both the dendrites and the general muscle membrane.[2] *Cambarus* apparently has only an accessory fibre innervating the dendritic region, and this supplies both RM_1 and RM_2.[195] Part of the motor supply is shared with neighbouring extensor muscles.

Physiology of the receptor organs. When the tension in the muscles of RM_1 and RM_2 is raised, the number of impulses passing towards the central nervous system along the axons of the sensory nerves is increased.[175, 176, 560] The two organs differ, however, in their type of response. RM_1 responds to continuous stretch by giving a rate of spike output which declines only slowly.[560] RM_2 adapts rapidly and after an initial burst of impulses is quiescent during continuous stretch. RM_2 also requires a greater degree of stretch than RM_1 to give any response at all. The threshold of response of RM_2 is thus greater than that of RM_1, and it is likely that RM_2 comes into operation only during the extreme flexion associated with swimming movements. It may, then, perhaps serve to induce reflex contraction of the extensor muscles.[560] The difference in threshold may be due in part to the mechanical properties of the two muscle organs, since the coefficient of elasticity of RM_2 is lower than that of RM_1. Hence a greater extension of the muscle is required in RM_2 than in RM_1 to produce the same tension in both.[308] The fast adaptation of RM_2 is unlikely, however, to be due to the mechanical properties of the organ.[308]

The rate of discharge in the nerve from RM_1 bears a linear relationship to the tension of the muscle,[308] suggesting that the critical factor involved in initiating sensory response is the degree of stretch imposed on the terminal endings.[176, 308] In further support of this conclusion is the fact that the sensory response is independent of the means by which the tension is raised. Thus, the sensory cells behave similarly, whether the tension is produced by passive stretching or by contraction of the muscle following stimulation of the motor nerve.[175, 560]

The responses of RM_1 and RM_2 to motor nerve stimulation differ. Stimulation of the efferent axon to RM_1 results in junctional potential changes at the muscle membrane similar to those at a vertebrate end-plate, the membrane potential being depressed below the usual resting value of 60–75 mV; but no spike is produced. Contraction of the muscle is slow, and is smooth at rates of nerve impulse above about 5 per sec. Stimulation of the axon to the fast adaptor RM_2, on the other hand, produces propagated spikes from the neuromuscular junctions; and the muscle contracts rapidly (twitch). The fusion frequency required to produce a smooth contraction is ten times as high as that for RM_1 i.e. 50 per sec.[310]   This difference in the mechanical response to motor stimulation should enable the animal to make sensory discrimination between tension due to passive extension and that due to muscle contraction.

Initiation of Sensory Nerve Response

It is not known precisely how increased tension in the receptor muscles is communicated to the sensory cell, the first recognizable step in the process being that the sensory cell dendrites are partially depolarized during stretch deformation. The normal resting potential of the sense cell, when the muscle is relaxed, is about 70–80 mV and this may be decreased by 10–20mV during stretch. The extent of the depolarization depends on the degree of stretch. This depolarization, termed the *generator potential*, spreads electronically and decrementally (i.e. is not propagated) from the dendrites and, if it still exceeds 8–12 mV in RM_1 sense cells or 18–20 mV in RM_2 sense cells when it reaches the cell body, a series of propagated impulses is initiated in the axon running to the central nervous system.[175]

Propagated spikes in the axon are probably initiated in the region of the cell body in the case of RM_2 sense cells but in RM_1 cells the point of origin is some 500 μ up the axon beyond the cell body.[169]   It is possible, however, that the precise point of origin may vary with the size of the generator potential.[174]

From observations of uninhibited responses of the sensory system, it would appear that the rate of discharge will be increased following contraction of the flexor muscles of the abdomen and contraction of the receptor muscles themselves. It will be decreased by contraction of the extensors in the preceding and succeeding segments, since this will tend to relax the muscle receptors.[560]   The situation is complicated by the fact that at least part of the motor supply to the muscle receptors (that to the tips of the muscles) is shared with the extensor muscles of the same segment,[2] so that, when these contract, the muscles of the receptor organ will also do so.   Analysis of the precise manner in which the receptor organs function is made still more difficult by the presence of a complex inhibitory system represented by the accessory nerves.

Inhibition

The process of inhibition has been studied mainly on crayfish where, it may be recalled, the single accessory nerve supplies only the region where the sensory dendrites end. In fact the inhibitory nerve terminations appear to synapse with the latter.

Stimulation of the accessory nerves prevents the initiation of a response to stretch by the sense cells, or results in a cessation of activity in the sensory nerve if it is already firing.[71, 312] The mode of action of the inhibitor appears to be similar to the inhibition of muscle contraction already described (p. 176 ff.). Thus activation of the inhibitor nerve results in little change, or a slight depolarization, of the membrane potential of the sense cell when the receptor muscle is fully relaxed.[312] However, if a generator potential is first produced by stretching the muscle, stimulation of the inhibitor nerve has the effect of restoring the membrane potential of the sense cell towards its resting level.[312] This repolarizing potential is termed the inhibitor potential. As inhibition thus tends to cause a slight depolarization in the resting membrane but to repolarize an active membrane, it follows that there must be an intermediate membrane potential at which stimulation of the inhibitor has no electrical effect. This membrane potential, known as the *reversal potential*, is found to lie within about 5 mV of the resting potential of the sense cell.[236, 312] Since the effect of inhibitor action is only to bring the membrane potential back to this reversal level, the size of the inhibitor potential produced by a given stimulus applied to the inhibitor nerve varies with the degree to which the membrane potential of the sense cell has previously been displaced from this level.[312] In other words, the inhibitor action acts like a voltage clamp tending to restore the membrane potential to a given level, whether it has been previously displaced by hyperpolarization or by depolarization. Since the reversal potential is close to the resting membrane potential, inhibitor activity of sufficient magnitude will prevent the spread of a generator potential to the soma following stretch of the sense cell dendrites.   Hence inhibitor activity can block sensory response to stretch.   Very similar effects have been observed during inhibition of the pace-maker potential of the vertebrate heart by vagus stimulation,[130, 271, 508] during inhibition of spinal motor

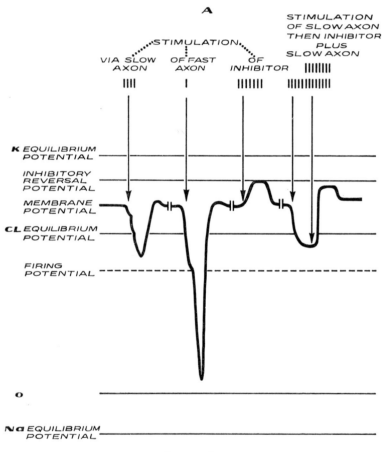

FIG. 54A.

neurones in the cat[108] and during inhibition at the crustacean neuromuscular junction.[180]

The effects observed during inhibition are comparable with those to be expected if the process brought about a marked increase in the permeability of the sensory cell membrane to inorganic ions whose equilibrium potential across the membrane lies close to the reversal potential.   When potassium is removed from the medium bathing the sensory cell, the resting membrane

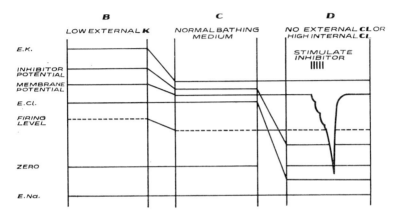

FIG. 54.   *A.* Diagram illustrating the relative positions of the various theoretical ionic equilibrium potentials, the resting and reversal potentials of the muscle membrane, and the effects of stimulation.   *B.* The effect of removing potassium from the medium.   The theoretical *K* equilibrium potential is raised, and the observed resting potential increases in magnitude.   The levels of the reversal potential and the firing potential are also raised.   *C.* The normal situation.   *D.* The effect of removing chloride from the medium or greatly increasing the internal chloride concentration of the cell.   The inhibitory reversal potential can be greatly decreased so that it is lower than the firing level.   Stimulation of the inhibitory nerve in these circumstances paradoxically results in a propagated action potential.   *EK* Potassium equilibrium potential; *E.Cl.* Chloride equilibrium potential; *E.Na.* Sodium equilibrium potential.   (*B.* and *C.* based on Edwards and Hagiwara,[167] *D.* Based on text descriptions.[236, 312])

potential is increased[167] (see also p. 172).   If potassium plays no part in the inhibitory process, the reversal potential might be expected to remain in its usual place after this change.   Hence, during the inhibition of a membrane depolarization of a given size, the inhibitor potential should be smaller than normal, or even of opposite sign (fig. 54).

In fact the effect of inhibitory stimulation in the absence of potassium is greater than normal on both the resting and active sense cell.[167]   It may be concluded, therefore, that the inhibitory effect is in part due to a great increase in the permeability of the sensory cell membrane to potassium, so that the membrane potential comes to be dominated more by the equilibrium potential of this ion than by those of other ions present.[167]   An alternative explanation of the effect might be that the permeability of the membrane to ions other than potassium is depressed during inhibition; but this seems to be ruled out by the observation that during inhibition the membrane conductance is much increased.[180, 312]

The chloride ion may also assist in the formation of inhibitor potentials.   Measurements of the inhibitory potentials by means of intracellular KCl-filled electrodes give different results from those where K_2SO_4 electrodes are used, suggesting that leakage of Cl from the former electrodes is sufficient to alter the Cl equilibrium potential across the membrane, and that this in turn affects the inhibitor potential.[236]   The inhibitor potentials are altered in size also when chloride in the medium is replaced by glutamate,[236] though such experiments are somewhat less conclusive, as some organic anions can themselves affect the potentials and so are not perfectly inert substitutes for chloride.[167, 311]   The present evidence seems to support the conclusion that in the stretch receptors inhibition is brought about by a large increase in the permeability to potassium and chloride ions clamping the membrane potential at near the resting level and so preventing the initiation and spread of the generator potential.

Central Nervous Connections of the
Sensory Supply from the Stretch Receptors

The sensory nerves from RM_1 and RM_2 run into the central nervous system via the second root nerve.   Somewhat curiously, however, the ganglion into which they run is the one in the segment anterior to that on which the stretch receptors are inserted. Thus the supply from the receptors spanning the intersegmental membrane of the fourth and fifth abdominal segments runs into the third abdominal ganglion.   In the ganglion the sensory nerves divide and, apparently without making any synaptic connections with interneurones in the ganglion itself, send off on the homolateral side of the nerve cord an ascending branch to the brain and a descending branch.   The descending branches from each nerve run as far as the sixth abdominal ganglion, where they all apparently synapse with a single interneurone, since the fibre of the latter responds to stretch in any segment (fig. 50).[268]

The muscle receptor organs are active during flexion of the abdomen.   No equivalent organ that responds to extension of the abdomen has yet been located anatomically.   Certain elements in the connectives between the abdominal ganglia, however, show increased activity when the abdomen is extended.   This activity is not abolished when the various sensory nerve roots are cut either singly or in combination suggesting that the response may be due to direct stretch excitation of the axons or of sensory receptors in the nerve cord sheath.[268]

Function of the Muscle Receptor Organs in the Control of Abdominal Posture

The extensor musculature of the abdomen of crayfish is divided into two morphological and functionally distinct regions.   The deeper layers of muscle are phasic fibres, responsible for twitch movements but incapable of maintained tonic contraction.   The superficial muscle fibres show tonic contractions and are those primarily used in determining the posture of the abdomen.[180b] One of the roles of the RM_1 organ is the reflex control of the activity of these postural extensor muscles.   Stretching of RM_1 results in activity in one of the five motor neurones to the superficial extensors. RM_1 activity never appears to influence either the other four motor neurones to the superficial extensors or the motor supply to the deep phasic muscles.[180b]   In the intact animal the discharge of the RM_1 organ is increased when the tail is flexed, and then progressively decreases as the abdomen returns towards the initial position.   Its activity, by stimulating extensor response, thus assists in counteracting passive change in the position of the abdomen once this has been pre-set.   Indeed the function of the organ appears to be related more to signalling changes in the position from a pre-set position than providing a measure of the absolute degree of flexion of the abdomen.   When active changes in abdominal posture are made by the tonic extensors there is a burst of activity in the RM_1 organs,[180b] presumably initiated by the contracture of the RM organ muscle fibre which shares a motor axon with the extensor muscles. This burst of activity reflexly excites the slow extensor fibres to further contraction which, since they are in parallel with the receptor organs, decreases the tension on the latter. The burst of RM_1 activity therefore

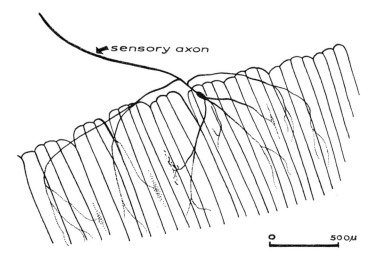

FIG. 55.   An N-cell from lobster muscle.   (From Alexandrowicz.[3])

declines and a new set position of the abdomen has been achieved. A necessary requirement for the operation of such a feedback system is that the motor supply to the RM should not be reflexly excited by the sensory output from the organ.   This is so.

Active flexion of the abdomen is accompanied by a burst of impulses in the inhibitory fibres to the muscle receptor organs.[162b] This serves to prevent a response from the RM_1 organs which would otherwise excite the extensor muscles and so oppose flexion.[180b]

2. ' N ' Cells

These are small sensory cells embedded in some of the thoracic muscles.   Unlike the situation in the RM_1 and RM_2 organs, these sense cells are not associated with specialized muscle cells, but instead terminate in long processes which ramify between the ordinary muscle fibres (fig. 55).

The ' N ' cells probably represent all that remains of a series of segmentally arranged muscle receptor organs whose function in the thorax of primitive malacostracans declined with the development of the carapace in decapods.[559b, 416b] This view is supported by the fact that the posterior thoracic segments of the stomatopod, *Squilla mantis*, contain muscle receptor organs but the homologous member of the series in the second thoracic segment is represented by an N cell only. This consists of a sensory axon which is sensitive to stretch and is slowly adapting. It is thus comparable with the RM_1 sense cell, which it may represent. The typical muscle and accessory nerves of muscle receptor organs are absent.[416b]

3. *Connective Tissue Stretch Receptors*

Limb Stretch Receptors

In the walking legs of decapods there is a strand of elastic tissue running between the dactylopodite and the apodeme of the closer muscle (the PD organ). In the strand lie the terminations of sensory nerves; and the structure serves as a proprioceptive organ responding to movements of the dactylopodite.[74, 78] Pairs of comparable structures give information about the movements of the carpopodite-propodite joint (CP_1 and CP_2) and also the meropodite-carpopodite joint (MC_1 and MC_2) (fig. 56B, D).[8, 553] A somewhat more complex organ spans the joint between the coxa and body wall.[10]

The PD Organ

In decapods, such as *Carcinus*, *Maia* and *Palinurus*, four types of sensory nerve fibre have been found running from the PD organ. Each fibre is responsible for mediating different types of information.[553] Two have cells close to the organ itself and continue to fire when the dactylopodite is either fully flexed or fully extended. The other two types, which have more centrally placed cell bodies, respond to movement associated with either flexion or extension respectively.   Information on both movement and position of the dactylopodite is thus signalled.[553, 555]   Sensory axons which respond in this way to uni-directional movements are not yet known outside the arthropods.   It is likely that the

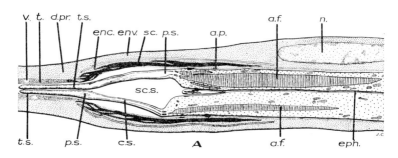

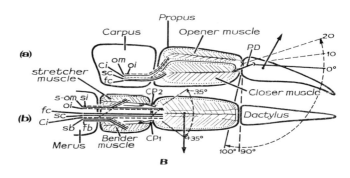

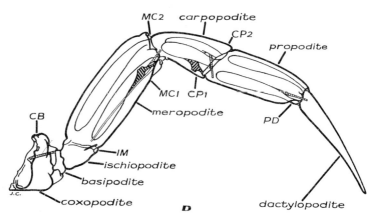

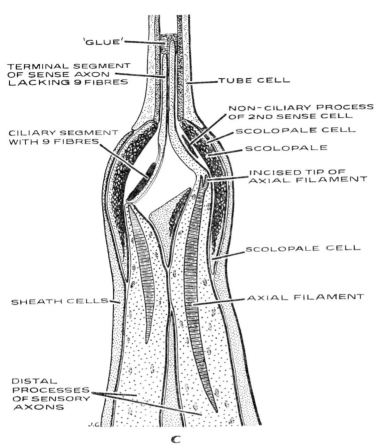

'GLUE'

TERMINAL SEGMENT
OF SENSE AXON
LACKING 9 FIBRES

TUBE CELL

NON-CILIARY PROCESS
OF 2ND SENSE CELL

SCOLOPALE CELL

SCOLOPALE

CILIARY SEGMENT
WITH 9 FIBRES

INCISED TIP OF
AXIAL FILAMENT

SCOLOPALE CELL

SHEATH CELLS

AXIAL FILAMENT

DISTAL
PROCESSES
OF SENSORY
AXONS

J.C.

C

FIG. 56. The stretch receptors of the limb. *A.* detailed structure of scolopidium from CP_1 organ: *a.f.* axial filament; *a.p.* attachment plaque; *c.s.* ciliary segment; *d.pr.* distal prolongation of scolopale cell; *enc.* scolopale cell enclave; *env.* enveloping part of scolopale cell; *eph.* ephase; *p.s.* paraciliary segment; *sc.* scolopale; *t.* tube; *t.s.* terminal segment; *v.* vacuole in wall of cell. *B.* Detail of the distal segments of the crab leg, showing the position of the PD and CP organs, the joint movements and the nerves involved in proprioceptive reflexes: *om* and *s-om* efferent axons; *oi* opener inhibitor; *si* stretch inhibitor; *Ci* common inhibitor; *sc* slow closer (motor); *fc* closer (motor); *sb* slow bender; *fb* fast bender. *C.* Diagram of scolopidium from PD organ. *D.* Generalized view of crab leg, showing the position of the various stretch receptors. (*A.* and *D.* from Whitear,[550a] *B.* from Bush,[78] *C.* from Whitear.[550])

proprioceptive organs are also capable of providing a measure of the rate of movement, since different fibres respond differentially to movement.[555]

The PD organ is responsible for initiating simple reflex responses which tend to antagonize passive movements of the dactylopodite. Thus, artificial ' opening ' of the dactylopodite is immediately followed by activity in the specific inhibitor to the opener muscle and also in the slow axon to the closer muscle.[78] Passive ' closure ' of the dactylopodite elicits discharge in the motor axon supplying the opener and stretcher muscles.[78] These effects are so specific and constant that it would seem that the reflexes may be of the simplest possible type (cf. fig. 47), involving a direct sensory to motor or inhibitor synapse in the central nervous system.[78]

The Carpopodite-propodite (CP_1 and CP_2) and Meropodite-carpopodite (MC_1 and MC_2) Proprioceptors

These organs consist of elastic strands containing sensory endings, CP_1 being similar to the PD organ while CP_2 is somewhat broader and flatter. CP_1 runs from the tendon of the bender muscle to the membrane of the joint; CP_2 is attached to the inner wall of the propodite and takes its origin on the tendon of the stretcher muscle.[553] Functionally the organs are distinct. CP_1 tends to respond mainly to extension, although showing some slight response to bending, whereas CP_2's response is almost entirely restricted to bending of the joint.[553] This is the opposite of what might have been expected, since it means that the sensory nerve activity is increased as tension in the elastic strand is decreased.[553] Detailed studies of the anatomy have been made in the attempt to throw some light on this point. Electron micrographs indicate that these receptors (and incidentally the PD organs also) are chordotonal organs similar in pattern to those which are widely distributed in the bodies of insects. Each of the elastic strands contains a number of embedded scolopidia in which the sensory axons terminate (fig. 56C). In the PD organ and CP_1 two sensory axons terminate in each scolopidium but they differ in the detail of their terminations. One axon has a segment comparable with a cilium in having nine peripheral filaments. (Since cilia themselves are of course absent in the Crustacea, the presence of this structure, apparently derived from a cilium, is of considerable phylogenetic interest.)   Such fibres are lacking in the other sensory endings.   The presence of two differing axon terminations might possibly explain the independent response to bending one way or the other in the receptors of the PD organ and CP_1.   The structure of a CP_1 scolopidium is indicated in fig. 56A.

The receptors MC_1 and MC_2 at the meropodite-carpopodite joint are essentially similar, in both gross structure and response, to CP_1 and CP_2.   Another comparable structure spans the coxopodite-basipodite joint.[8, 78b]

The CP organs initiate reflex responses antagonizing passive joint movements in a way similar to that described for the PD organ.   Thus, passive stretching (reduction of the propopodite) is followed by activation of the slow motor axon to the bender; bending (production of the propodite) elicits activity in the slow motor axon to the stretcher.   Activity in the respective fast motor axons is also produced if joint movements are made very rapidly.[556]   Both position and movement receptors are present, therefore, as in the PD organ.

It must be stressed that all the activity so far studied in the PD, CP and MC organs has consisted of responses to passive movements of the joints.   The responses of the organs to movements produced by muscle activity are not known, though they would be expected to be similar.

The reflex responses of the different joints are not entirely independent, since the position of the propodite has been found to influence the efferent axon response to movement of the dactylopodite.[556]   Position of the dactylopodite, on the other hand, does not affect the responses of the more proximal joints.[556]   In addition to this interaction of sensory information from more than one joint affecting the motor response within a single limb, there are also indications of a wide spread of sensory response in the central nervous system.   Thus, when the legs of crayfish are passively extended or flexed, reflex discharges are produced in the leg nerves of other limbs, particularly of the homolateral side, but with some spread to the contralateral side.[423]

Movement is possible in more planes at the joint between the coxa and the thorax than at the internal limb joints, and in consequence the former is equipped with a more complex set of proprioceptive organs. Three types of structure are present (1) a muscular receptor across the thoracico-coxal articulation, (2) a connective tissue receptor spanning the same joint, and (3) two innervated elastic strands lying alongside the levator and depressor basipodite muscles respectively. (In addition there is an elastic receptor comparable with the more distal elastic receptors across the coxo-basipodite joint.) These organs are not necessarily all present in any given species. *Homarus* has a full complement but *Eupagurus* lacks the connective tissue receptor. The development of the levator and depressor receptors seem to be related to the degree of agility of the species. They are well developed in *Eupagurus* and *Carcinus*, small in *Cancer*, and apparently absent in the sluggish *Maia*.[8]

None of these structures has yet been physiologically studied, but their position provides evidence as to their probable functions. In *Homarus* the muscle fibre probably contracts when the leg is moved forward, whilst the elastic receptor is stretched by rearward movement of the leg. These two receptors are so located, therefore, that they are in a position to provide information concerning the movements of the leg in the horizontal plane.[8] When only the muscle organ is present, as in *Eupagurus* and *Carcinus*, the two sets of sensory nerve endings are arranged in a manner which implies that one responds when the muscle contracts and the other when its surrounding strands of connective tissue are stretched (as by backward movements of the legs). The one organ in these animals could therefore combine both functions.[8]

Neither the muscle organ nor the elastic strand at this joint are chordotonal organs.[550]

The levator strand is extended by the contraction of the depressor muscles pulling the leg downwards. It is thus likely to come into action when the animal stands up and when it walks.

The depressor strand lying on the depressor muscle is extended if the basipodite is pulled upwards. These two organs, therefore, probably respond to movements in the vertical plane.

Activity can be detected in the central nervous system following movements of many other joints; which suggests that proprioceptors are present at sites where they have not yet been located anatomically.[556] Proprioceptive impulses from the swimmerets, of crayfish for example, seem to be essential to the control of the rhythm of the beat,[267] although the organs themselves are still unknown.

Joint receptors comparable with those of the P.D. organ have been found in the three distal joints of the antennule of *Panulirus argus*.[561b]

2. Exoreceptors

These include the organs responsible for the senses of sight, touch, balance, chemo-reception and (probably) pressure sense.

TACTILE SENSE

Three distinct types of sensory information are provided by various superficial sensory structures on the appendages and body of the crustacean. These include contact sense, detection of speed and direction of water movements over the body surface, and detection of distant disturbances in the water. Apart from the PD organs, which respond to low-frequency vibrations transmitted to the dactylus via the substrate,[74] the organs responsible for tactile sense are all modified hair sensillae.

Many of the hairs covering the appendages are fairly stiff along their shaft but bend at the base. Such bending activates neurones situated at the bottom of the shaft (usually two in number in crustacea, but only one mechano-receptor in the insects). When the hairs are touched spike discharges are initiated in the sensory axons which lead to the central nervous system. Quite a heavy force is required to displace robust hairs, such as those lining the cutting edge of the claw or the guard hairs on the antennules. The sense organs of such hairs respond only to mechanical movement, and not to water disturbances. Hairs of this type are also common along the edge of the swimmerets and dorsal surface of the body in crayfish; and the ways in which tactile information derived from them is interpreted in the central nervous system have already been discussed (p. 195 ff.).

Two organs are known which detect movements of water past the body surface. These are the pit hair receptors of the thorax of *Procambarus* and the hair peg organs of *Homarus*. The former, which are widely distributed over the carapace, occur in shallow depressions in the cuticle (fig. 57*A*).   The functional unit is a tactile hair some 300 μ long, at whose base are two neurones which send dendrites up the shaft of the hair.   Associated with the tactile hair are two companion hairs which are not innervated.

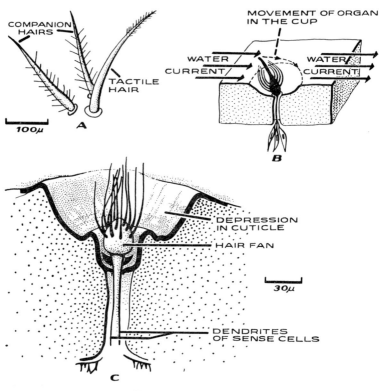

FIG. 57 *A*, *B*, *and* *C*.

Deflection of the tactile hair one way causes one of the two neurones to initiate nerve impulses in its axon, deflection in the opposite direction excites the other.   The faster the hair is deflected, the greater the initial rate of firing in the sensory axon.   Adaptation is comparatively rapid if the hair remains deflected, the cell returning to its resting rate of firing in about 50–60 seconds.[375] From these properties the receptors may be presumed to be capable of signalling the direction and speed of sudden movements across the body surface but not of recording sustained, steady currents.

The manner in which the neurone is excited is unknown but one suggestion is that the neurone and its dendrites are stretch receptors which fire when mechanically distorted. If the arrangements of the neurones were as in fig. 57D then one neurone's

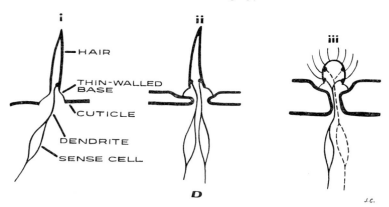

FIG. 57. Various types of tactile receptors. *A*. A pit receptor composed of a sensory hair (right) flanked by two non-sensory 'companion' hairs. *B*. The hair peg organ, showing how it is deflected within its cup by a stream of water passing over the surface of the cuticle. *C*. The hair fan organ. *D*. Possible types of innervation in crustacean mechanoreceptors: (i) occurs in the statocyst hairs having only a single nerve fibre, (ii) the situation reported for *Procambarus* with two dendrites inserted on opposite sides, (iii) postulated situation in *Homarus* hair fan with two nerve units innervating the base at different sites, both responding to deflection in two directions but with different sensitivity. (*A*. from Mellon,[375] *B*. from Laverack,[326] *C*. from Laverack,[327] *D*. from Laverack.[329])

dendrite would be stretched and the other's relaxed when the hair was displaced towards the right, and vice versa for leftwards displacement. If so this would provide an explanation of the differential response of the two neurones.[375]

The hair peg organs of *Homarus* (fig. 57B), which are widely distributed over the front half of the body and on the legs, are a little more complex in structure. Each lies in a shallow depression and consists of a short peg-like hair from which arise numerous fine hairs.   The peg-hair is hinged at its base so that it has a greater freedom of movement in one plane than in the other.   At its base, $3-5$ sense cells send dendrites into the shaft.   At least one of the sense cells is mechanically activated when the peg organ is displaced.   The threshold current speed required for neurone response is 0.25 cm per sec., and the hinged arrangement of the peg hair provides a potential means by which central integration of the output of a large number of such units (and there are as many as 75 to the square cm on lobster chelae) may make possible the detection of the direction of current flow.[326]   The sense cells adapt only slowly, making continuous monitoring of the current possible.   It is probable that only some of the sense cells at the hair base are mechano-receptors; the others may include chemo-receptive or other sensory components.

At moult, the hair peg organs are lost; and, as new ones are not obvious in immediately post-moult animals, they must be secreted subsequently.[326]   In consequence the ability to detect currents seems likely to be much diminished at this time.

Rather similar structures—the hair fan organs (fig. $57C$)—are responsible for detection of agitation in the water.   The central hair of these hair fan organs is somewhat flattened and lacks the peg of the peg organs.   Like the latter, however, it is hinged so that only in one plane is it readily displaced.   The organs are widely distributed over the front half of the body in *Homarus*.

Disturbance in the water causes displacement of the hair fan, and this is signalled to the central nervous system.   Adaptation of the sense cells to steady currents is rapid, though firing is continuous if the water nearby is agitated.   The structure is therefore presumed to detect disturbance in the water, such as might be caused by the approach of another animal, rather than steady currents.[327]

The hair fan organ has two neurones at its base but both these appear to be active whichever way the hair fan is deflected —unlike the situation in the pit hair organ of the crayfish.   The mode of dendrite ending in the hair fan may differ, therefore, from that of the latter.   Three types of innervation of various tactile sense organs have been postulated; these are illustrated below.   Clearly only in figure $57D_2$ can displacement to both left and right be differentially signalled.   This is the arrangement postulated for the pit hair of *Procambarus*, the only tactile receptor where this seems to be possible.

Pressure, Balance and Detection of Movement

Two types of sensory information desirable in motile forms are those of balance and acceleration. An animal must know which way up it is with respect to gravity, and also have information concerning its rate of movement, particularly if angular displacements are occurring.

Specific organs (statocysts) responsible for provision of such sensory information are known only for the Malacostraca. In lower forms orientation is primarily with respect to light rather than to gravity. This can be shown in *Artemia* by covering their eyes and illuminating from below, whereupon the animals will swim dorsal side uppermost instead of in the usual inverted condition. A comparable dorsal light reflex can be elicited in mysids after removal of the statocysts.[441]

Well developed statocysts are present in most decapods and mysids, but only in some species of other malacostracan groups. The site of the organ varies, being in the endopodite of the uropods in mysids and in the basal joint of the antennule of decapods. The statocyst arises as a result of the invagination of the body surface, the lumen of the capsule so formed containing hairs comparable with those used in the touch senses of the general body surface. A statolith also is usually present. The hairs are often specialized in structure, two types being present in the lobster. The thicker ones are associated with the statolith (sand grains cemented together) whilst the longer, thinner hairs are free (fig. 58). In mysids there is only one statolith and this is secreted by the cells of the organ.

Statoliths, being dense bodies, tend to displace their associated sensory hairs in a direction dictated by the position of the animal with respect to gravity. The sense cells associated with these hairs are potentially suited to signal the orientation of the animal with respect to gravity. That they actually perform this function was clearly demonstrated in the classical experiments of Kreidl.[306] Relying on the fact that the whole of the superficial material of the statocyst (internal cuticular lining, hair cuticle and statolith) is lost at moult and the statolith replaced from the environment, he substituted iron filings for sand in the aquarium. His experimental prawns (*Palaemon*) duly utilized the iron filings in the

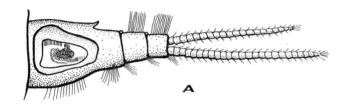

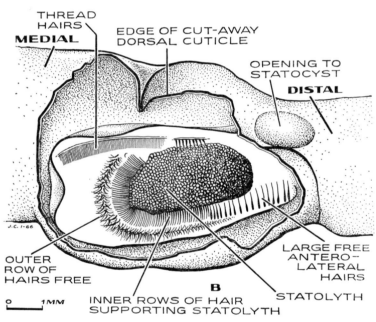

FIG. 58. *A.* Basal segment of antennule opened to show the statocyst. *B.* Enlargement of the statocyst, showing sensory hairs and statolyth. (Redrawn from Cohen.[105])

statocyst and were then shown to orientate themselves in a magnetic field in a manner indicating that their position was determined by the resultant of the gravitational and magnetic forces acting on the artificial statoliths.   As a result of the loss of the lining of the statocyst at moult, some swimming crabs, such as *Platyonichus ocellatus*, may at this time temporarily lose their sense of balance and roll forward round the transverse axis.[104]

The sense cells signal to the central nervous system continuously and, if the signals indicate that the animal is no longer in a horizontal plane, reflex correction movements are initiated.   Thus, if an animal is rolling to the right, the left hand legs are flexed and raised whilst the right hand legs are extended and depressed, a movement which would tend to restore equilibrium.   That the statocyst initiates this correction can be shown by displacing the hairs of one statocyst with a jet of water, whereupon the animal performs the reflex act even though the body remains in an even plane throughout.

When the animal is in the normal upright position, the signals from the two statocysts are cancelled centrally.   Removal of one statocyst results in the animal leaning over initially towards the affected side, though after a few days this becomes less marked as a result of central nervous compensation.

Investigation of the activity in the axons of the nerve leading from the statocyst of *Homarus* indicates that four types of receptor are present: vibration, acceleration, and two types of position detectors.   Sense cells associated with the shorter hairs are responsible for position responses and cells at the base of the thread hairs respond to acceleration.   The type I position receptors respond to movement by a non-adapting discharge, the frequency of which is related to the degree of displacement about the transverse axis from the normal stance (fig. 59).   These receptors respond only to movement in the transverse plane.   It may be noted that if the animal had only one such receptor it would be unable to determine whether rotation was tail down or tail up, as the response curve is dome-shaped (fig. 59).   This factor is accounted for by the considerable number of receptors in each statocyst, the response curves of which differ slightly, thus making it possible for integration within the central nervous system to determine the direction of movement.

Type II position receptors signal movement either about the longitudinal or the transverse axis, though no one receptor performs in both planes. These receptors can also signal information concerning the direction of movement as well as position.[105]

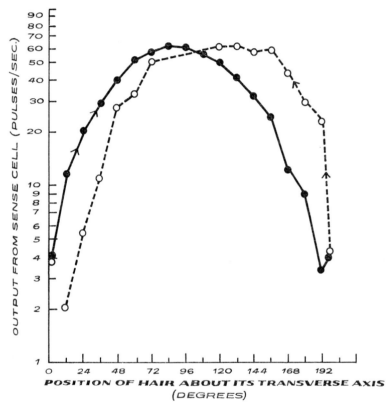

FIG. 59. Response of a single statocyst position receptor to rotation of the animal about its transverse axis. ○ - - - ○ - - - Rotation head downwards, ● - - - ● - - - Rotation head upwards. (From Cohen.[105])

The Detection of Acceleration

The reader will doubtless remember from his childhood the after-effect of spinning around until dizzy. Rotation results in the forcing of fluid through one or more of the semi-circular canals and this continues for a time after the body comes to rest. One reflex result of the fluid movement associated with this response to angular deflection is nystagmatic movements of the eyes giving the subjective impression that the environment is rotating past the observer. A crab exposed to rotation also shows nystagmus, indicating the presence of sense organs capable of detecting angular movements. Denervation of a row of thread-like hairs in the statocyst abolishes the response.[141] Fluid movements produced by rotational forces would deflect these hairs. Records of the sensory output of the thread receptors indicate that they respond only to movement, and indeed that they adapt quickly if the rate of movement is kept constant.[105] Acceleration is therefore the factor primarily determined. Differential responses allow for the detection of angular acceleration in all three planes.

To sum up, therefore, the decapod statocyst is a receptor system able to detect (1) body orientation with respect to gravity, (2) direction of movement about the longitudinal and transverse axes, and (3) detection of acceleration in three planes. The overall capabilities of the statocyst as a balance and movement-detecting organ are thus comparable with the vertebrate middle ear. It is not surprising, therefore, that removal of both statocysts usually results in disorientation of the animal. Thus if *Uca* (*Gelasimus*) *pugilator* loses both inner antennae, it tends to roll over backwards when running. The crab tries to right itself after inversion but often with such uncontrolled violence that it rolls forward again right over onto its back.[104] Such behaviour clearly indicates that its senses both of position and of acceleration have been lost. In mysids bilateral extirpation of the uropods, though it completely eliminates the statocysts, does not prevent the animal from swimming dorsal side uppermost even in complete darkness. How this orientation is achieved is uncertain but one suggestion has been that the gravitational displacement of the inner organs may be detectable by internal sense organs.[441]

Pressure

Many marine species, particularly those planktonic forms which undergo diurnal vertical migrations, need to be able to monitor

16

their depth in the water if they are not to risk sinking out of their feeding zone.   On a bright day the intensity of penetrating light could be utilized, but it is apparent that any form solely dependent on phototaxis would be liable to locate itself at an incorrect depth as a result of differences in illumination on bright and cloudy days. Another means of gauging depth is by sensing the hydrostatic pressure, and it appears that a number of Crustacea can do this.

Increase in pressure usually elicits an increase in swimming movements, whereas a decrease in pressure is followed by a decline in activity.[442]   The response is non-directional and hence appropriate orientation can be brought about only by the intervention of other sensory information.   Light and gravity seem to be the most important orientating agents.

The planktonic crustacea so far studied can be classified in three groups according to their responses to these factors[442]:

Type 1.   The direction of swimming is orientated entirely with respect to gravity, pressure increase initiating upward swimming and decrease in pressure resulting in passive sinking irrespective of the direction of the incident light.   *Example*: the megalopa of *Carcinus*.

Type 2.   Movements are orientated with respect to gravity when the light source is vertically above or below the animal. These forms move towards a horizontal light when pressure is increased and away from it when it is decreased.   *Examples*: *Praunus*, *Schizomysis*, *Siriella* and *Leptomysis*.

Type 3.   Pressure increase enhances movement towards the light source in whatever direction this may lie.   Various types of response follow decrease in pressure:

(*a*) Pressure decrease does not result in active swimming away from the light.   Thus, if the light is from below, the animals tend to accumulate on the bottom.   *Examples*: *Hyperia* (Amphipoda), zoeae of *Nephrops*, *Pandalus*, *Crangon*, *Galathea*, *Pagurus*, *Carcinus*, *Cancer*.

(*b*) Pressure decrease results in active swimming away from the light source.   *Examples*: nauplii of *Balanus*, the larvae of the euphasid *Meganyctiphanes*, and the copepod *Temora*.

(*c*) Pressure decrease is followed by active movements away from the light, passive sinking, or a mixture of both.   *Examples*: the copepods *Acartia* and *Calanus*.

The separation of the responses into the three primary group-ings can be related to the degree of development of statocysts. Forms which lack statocysts, e.g. the type 3 animals, respond to light, from whatever direction it is incident.   On the acquisition of a statocyst at the moult to the megalopa however, the gravity response becomes dominant, e.g. the type 1 *Carcinus* megalopa.

A few animals, of which the amphipod *Caprella* is one, respond to increased pressure by less active swimming and to decreased pressure by more active swimming.   They thus tend to sink when the pressure is increased.   Clearly such behaviour would be catastrophic to a planktonic form, but *Caprella* is a littoral and sub-littoral form.   To an animal from this habitat this ' reversed ' response to pressure may be advantageous in decreasing the risk of stranding at low tide.[442]

Yet another type of response is found in the amphipod *Synchelidium*.   This animal lives and feeds in the wave stirred zone of Californian beaches, moving up and down with the tide. Its behaviour seems to be adapted to avoiding the obvious risk of being stranded.   When the tide is rising, a sudden increase in pressure—such as might result from a wave passing overhead— causes a burst of activity, which declines again after a few seconds. The bigger the pressure increase, the longer the activity persists. The animals therefore tend to be carried up the beach, but not to the ultimate limit reached by the wave.   When the tide begins to ebb, the length of the swimming period following any given pressure increase is lengthened.[172]   Presumably this ensures that the animals will continue swimming as a wave recedes and so decrease the chance of being stranded (fig. 60).

Gas bubbles are apparently absent from the crustaceans show-ing pressure responses and hence, although the sense organs responsible for detecting change are not known, it is unlikely that they operate by detection of gross volume change.   Detection of change in the volume of macro-molecules is one suggestion[172] as to the method used; but, as behavioural responses can follow pressure changes as small as 15 cm of water, this would imply that the capacity to respond to volume changes is of the order of 1 part in 12,800,000.   Such a high order of sensitivity does not seem very likely.   An alternative scheme for pressure detection is based on the observation that when a copper electrode is inserted into the abdomen of a prawn and is connected via a galvanometer to another electrode in the medium, the current flow is sensitive to pressure.[140] This effect is comparable to a similar pressure

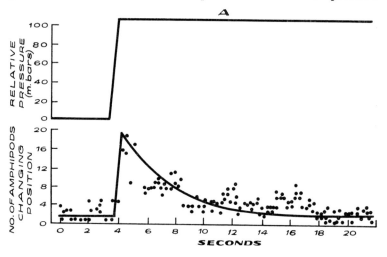

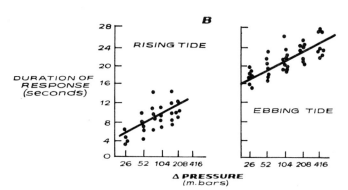

FIG. 60. Activity responses of the amphipod, *Synchelidium*, to sudden changes in pressure. *A*. Response to 110 mbar pressure change. *B*. Differences in the magnitude of response to pressure change at various stages of the tidal cycle. (From Enright.[172])

sensitivity observed in an artificial arrangement in which two concentration cells dipping into saline are connected via a galvanometer to a reference electrode in the saline solution. A further similarity between the animal and the artificial system is that in both cases the pressure sensitivity is destroyed by the presence of a large cation such as that of Cetavlon, but not by large anions such as those of Teepol. The pressure sensitivity of both artificial systems and animals is restored by rinsing in dilute H_2CO_3; which suggests that the area sensitive to Cetavlon is at the surface.[140] The sense organs involved in pressure detection, however, remain unknown.

SIGHT

Sight is perhaps the most important exoreceptive sense in active animals and, with the exception of a few subterranean forms living in complete darkness—such as the blind white crayfish of the river Styx in Mammoth Cave, U.S.A., and some interstitial forms—most crustacea have functional eyes during at least some stage in their life cycle. In this group eyes show a wide spectrum of structure, ranging from the simple median eye of nauplius larvae through the more advanced median eye of copepods to the complex compound eyes of the adults of most other groups. The last reach their greatest degree of development in the Malacostraca. There are also indications that in addition to the eyes, at least some parts of the body surface of a few forms are also sensitive to light. Thus, a number of the lower crustacea can still display the dorsal light reflex even when the eyes have been covered or removed. The mechanism of the action in this case is not known; but in the crayfish the last abdominal ganglion has been found to be sensitive to light,[289, 543] both in epigean forms and in eyeless cave species.[325b] Chromatophores in many species will also respond to changes in light intensity after the eyes have been covered.

The Median Eye

Median eyes are present in the larval and adult stages of most lower crustacea and also in some Malacostraca. In general the median eye consists of a small group of ocelli, two of which are disposed laterally and usually one or two medially. Each ocellus is composed of a limited number of light-sensitive cells, from which fibres lead to the brain from the peripheral side as in the vertebrate eye.   Median to the light-sensitive cells there is usually a reflecting layer of pigment which prevents the entry of light from the side and from behind.   A lens focusing light onto the sense cells is often present, as in *Artemia*, but is usually absent in copepods.

Relatively little is known of the function of the median eye; but it may be presumed to enable the animal to orientate to light, as is necessary, for example, in the case of the diurnal vertical migration of copepods.   Other functions are possible, such as the detection of movement or of the plane of polarization of light. The use of the eye for discriminating static features of the environment by image formation seems unlikely in view of the very limited number of sense cells in each ocellus.

The Compound Eye

As the name suggests, the compound eye is composed of a large number (some 13,000 in *Astacus*) of sub-units or ommatidia. Each ommatidium (fig. 61) consists of a number of secreted regions in addition to the living cells.   The former, which pick up the light path leading to the light-sensitive cells, consist of the outer corneal lens, a crystalline cone—which in some eyes is extended into a crystalline tract—and finally a crystalline rhabdome.   The cellular components comprise modified epithelial cells which are responsible for secreting the corneal lens, cells which secrete the crystalline cone and are responsible for its enlargement during the growth of the animal and a group of retinulae, the light-sensitive cells which surround and secrete the rhabdomeres which comprise the rhabdome.   In addition, a number of pigment cells also are present.   At moult, the corneal lenses are lost and replaced by the secretion of the modified epithelial cells.   The arrangement of these various components of the eye is not constant in all forms; two general types of eye are found.   These have been termed the *apposition* eye and the *superposition* eye, in reference to assumptions made as to the manner in which images are focused on the sense cells.

Apposition Eye

The apposition eye is found principally in forms exposed to relatively high light intensities, e.g. terrestrial species such as

Grapsus grapsus,[536] diurnal littoral forms such as *Eupagurus bernhardus*[52] and the mangrove crab *Goniopsis cruentatus*.

In each ommatidium of these types of eye the retinulae extend from the basement membrane to the bottom of the crystalline cone. The latter is surrounded by the distal pigment cells

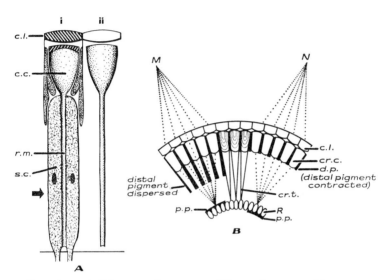

FIG. 61. *A*. Diagram of an ommatidium of an apposition eye: *c.c.* corneal lens, *s.c.* sense cells. In the transverse section the individual rhabdomeres can be recognized in the rhabdome (*r.m.*). *B*. Diagram of the paths of light rays in the superposition eye. The distal and proximal pigment cells are shown in the light-adapted state at the left of the Figure, and in the dark-adapted state at the right. The centre three ommatidia are drawn to show the laminated appearance of the cones and the position of the crystalline tracts. On the diagram: *R* retinule, *c.l.* corneal lenses, *cr.c.* crystalline cone, *d.p.* distal pigment, *p.p.* proximal pigment, *cr.t.* crystalline tract, *M* and *N* sources of light. (From Kuiper.[314])

which prevent light reflected from neighbouring ommatidia from entering the light path (fig. 61). As the retinulae themselves also contain dark pigment, which must similarly screen out reflected light, it is probable that excitation of the sense cells is solely by light entering the corneal facet of that ommatidium and passing via the cornea and the crystalline lens to the rhabdomeres, where it is thought that the initial stage of the light response occurs. If the length of the light path is about the same as the focal distance of the lens, then an inverted image would be in focus at the level of the rhabdome. It is very doubtful, however, whether the individual ommatidium is capable of receiving such an image, since there are only 7 (sometimes 8) retinulae cells clustering round the rhabdomes. It is presumed, therefore, that the output of nervous activity from each ommatidium is related to the light intensity falling on it. This in turn will be related to the amount of light reflected from objects within the narrow field of view of any one ommatidium. As the facets of each ommatidium in the eye are at a slight angle to one another, each will have a different field of view. The eye has therefore been presumed to build up a mosaic picture of the environment as a series of dots of different shades, the intensity of each dot representing the signal intensity from any given ommatidium. A newspaper picture is formed in a comparable manner, by a large number of equally spaced black dots, though here areas of different intensities are produced by varying the size of individual dots.

The building up of the overall image in this manner depends essentially on there being but little overlap in the fields of view of different ommatidia so that only light rays travelling parallel to the axis of the ommatidia are received. This might well be the case if there were little difference in the refractive index of the optical light path of an ommatidium and the surrounding tissues, so that any light waves entering the ommatidium at an oblique angle could readily pass out again without stimulating the sense cells. However, it appears that the refractive index of the light path is considerably greater than that of the surrounding tissues, with the result that light entering the ommatidium tends to be trapped so that a high proportion, even from an angle as great as 10° either side of the longitudinal axis, reaches the sense cells.[314] This means that there is bound to be a measure of overlap in the fields of view of neighbouring ommatidia; this in turn would be expected to decrease the sharpness of image formation by the eye as a whole. Part of a possible solution to this difficulty may be that the different sense cells within an ommatidium have smaller fields of view than the whole, by virtue of slight angular displacements of the rhab- domeres individually associated with each sense cell. When a transverse section of the rhabdome is studied while a tiny beam of light is passed across the surface of the eye, it has been found[314] that each rhabdomere can be made to light up in turn. The field of vision of each rhabdomere is only some $3-4°$.[314] Nevertheless, even assuming that a sense cell will respond only to light passing in its own rhabdomere—of which there is yet no proof—there will still be some overlap of visual field both of the sense cells within an ommatidium and of neighbouring ommatidia. Thus, although the ommatidium is the morphological unit, it is unlikely to be the functional unit. Discrimination must therefore involve integration by the central nervous system of the output from neighbouring sense cells. It is interesting in this respect that the nerve fibres from neighbouring ommatidia are linked (fig. 62).[314] Precisely how integration is achieved by the central nervous system is not yet fully understood, but investigation of the inter-neurones found in the optic tract indicates that already at this level fibres can be found which are responding to and presumably also integrating the output of a considerable number of ommatidia (fig. 62).

Superposition Eye

This type of eye is found chiefly in nocturnal or sub-littoral forms such as *Palinurus, Galathea, Dromia, Portunus,* and *Pisa,* though also in some forms from shallow water such as *Leander* and *Palaemonetes.* Structurally the eye differs from that of the apposition eye principally in that the crystalline cone is drawn out proximally to a considerable extent to form a crystalline tract. The retinulae cells are not therefore in direct contact with the crystalline cone itself, but are at a considerable distance from it. The distal pigment cells surround the light path of the ommatidium like a jacket, and they contain pigment which can be either dispersed so that the ommatidium acts as a single light path, or concentrated so that light can enter after being reflected from neighbouring ommatidia (fig. 61B). This led Exner (1891) to suppose that it was possible for this type of eye to function in two different ways. If the pigment in the distal cells was dispersed, the eye was assumed to behave in a manner similar to the apposition eye, each ommatidium conducting only light waves

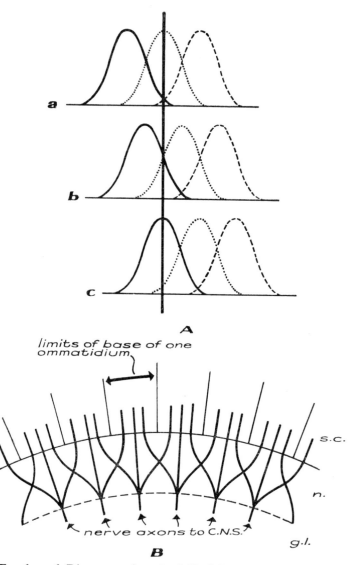

FIG. 62.   *A*. Diagram to show the field of three sense cells of three adjacent ommatidia, *a*, *b* and *c*.   The vertical line indicates a point in the field of vision of the three ommatidia.   *B*. Diagram of the pathways of the sensory nerve fibres (*n*) from the sense cells (*s.c.*) towards the ganglion laminaris (*g.l.*).   (From Kuiper.[314])

running parallel with its axis. If, however, the pigment were concentrated, then light from a single source could be focused by a number of ommatidia to fall on the sense cells of one (fig. 61B). The light-gathering powers of the eye would be much increased by such a system and this would therefore be advantageous in a dim light. Some loss of visual acuity would be expected as a result of the multiple passage of light through different lens systems. Vision in bright conditions would therefore produce a mosaic picture with reasonably sharp image, whilst in a dim light the concentration of the pigment would allow a greater proportion of the light incident on the eye as a whole to reach the sense cells, but with some loss of resolution. This theory, with slight modifications, has been generally accepted; but it is doubtful if it can explain certain recent experimental observations. One of the requirements for Exner's theory is that the crystalline cone should have a higher refractive index at the centre than at the periphery. Examination of the cone in the presence of liquids of different refractive index does not indicate that this is the case.[125] Furthermore, a consequence of the operation of the eye in the dark-adapted state (i.e. with pigment contracted) would be that the image formed would be an erect one. Microscopic examination of the image formed behind the cones in various forms including *Palinurus, Galathea* and *Dromia* indicate that the image is, in fact, inverted.[314] In addition, beams of light originating at a small source and passing through the lens system converge, not at the level of the retinulae cells, but at a point considerably behind them.[314] It seems likely, therefore, that in these crustacea the eye does not work on the superposition principle. Nevertheless, in some insects the position images are formed at the level of the rhabdome. Indeed in forms such as the locust a whole series of images can be detected if a microscope is focused at different levels in the rhabdome.[76] The most distal of these is the image produced by light beams entering the same ommatidium. The proximal beams are produced in series by refraction and reflection from progressively more distant surrounding ommatidia. Still more distant ommatidia focus images which, like those in the crustaceans mentioned, fall behind the rhabdome and so are not detectable. Contrary to what might have been expected, the resolving power of the more proximal images is greater than that produced by the focusing apparatus of the single ommatidium. The latter can produce an image in which points subtending an angle of about 1° at the ommatidium are separable, whilst a deep image can resolve points subtending only 0·3°. An indication that movements of this latter order of magnitude can actually be detected is given by the fact that an angular displacement of 0·3° of a small light source results in nervous activity extending as far as the ventral nerve cord.[76]

Prawns and the mysid *Praunus* do not respond to objects subtending an angle of less than 4·6° at the ommatidium[125]; which perhaps offers further confirmation that their eyes do not form usable superposition images. It must be remembered, however, that values for visual acuity determined by studying changes in behaviour do not necessarily delimit the optical capacities of the sense organ itself, but may involve components of the central nervous system. Thus, to excite the central nervous system to initiate behavioural responses may need a bigger change than the minimum detectable optical response.

The fact that prawns and mysids do not appear to use the superposition principle does not, of course, necessarily imply that no crustacean can do so.

Pigment migration in the superposition eye

Three types of pigment cells are usually present; two of these contain a black pigment, ommatin, related to melanin, and the third contains a white reflecting pigment. Black pigment is present proximally in the seven or eight retinulae cells themselves, and peripherally in two distal pigment cells which surround the crystalline cone and tract. The white pigment is contained in cells lying at the level of the basement membrane. Processes of these cells extend on either side of the basement membrane.

The pigment in all three types of cells can be moved to adapt the eye to the prevailing light conditions. In eyes adapted to high light intensity, the pigment in the white reflecting cells is withdrawn through the basement membrane and its reflecting power is correspondingly diminished. The black pigment in both the proximal and distal cells is dispersed, and so forms a screen isolating the ommatidium from its neighbours (fig. 61B). Conversely, in dark-adapted eyes the distal and proximal pigments are concentrated, and the white reflecting pigment migrates above the level of the basement membrane, thereby increasing its efficiency.

These movements of pigment have been used as indirect evidence in support of a theory that superposition images are used by crustacea. The argument runs as follows. In the light-adapted eye the ommatidia can act separately forming a mosaic image; but concentration of the pigment in the dark-adapted state must be for the purpose of increasing the light gathering power of the eye by using a number of ommatidia to form super-position images. Were it not that the optical evidence already cited seems to preclude the superposition theory in a number of cases, this would seem a reasonable assumption. Any counter-proposals must indeed provide a satisfactory explanation of two facts, (a) that the pigments move during light- and dark-adapta-tion, and (b) that there is a long distance between the crystalline cone and sense cells in the eye of the anatomical superposition type. Recently, Kuiper[314] has put forward an alternative theory which can explain these two points. He assumes—contrary to the superposition theory—that it is not low, but high, light intensities that are critical to the crustacean eye. The pigment movements are understood to have a protective role at high light intensities. In support of this, it is worth noting that dark-adapted crustaceans with the superposition type of eye display evident signs of alarm when placed suddenly in a bright light, whilst those with the apposition type of eye do not. Kuiper argues that the pigments may function in the following manner. The crystalline tract and cone have a high refractive index and, when no pigment surrounds them, can act as a light guide, trans-mitting a high proportion of the light incident on the ommatidium facet to the rhabdome. The pigment also has a high refractive index and hence, if placed in close contact with the re-flecting surface of the cone and crystalline tract, will allow the escape of some light rays into the pigment cell, where it is likely to be absorbed. The longer the light path, the greater will be the proportion of the incident light that is potentially capable of being absorbed in this way and the greater the protection obtained from high light intensities. This is the reason for a wide separa-tion of cone and sense cells in this type of eye.

The proximal and distal pigment cells do not behave in an identical fashion during light- or dark-adaptation. In prawns, the time taken for the pigment to migrate from the fully dark-adapted to fully light-adapted state takes some 40–90 minutes for the distal pigment but only some 4–6 minutes for the proximal pigment. The proximal pigment movement is often complete, therefore, before distal movement has begun.

The distal pigment undergoes cyclical dispersion and concentration, independently of the light intensity. Thus, the eye tends to be in the concentrated (dark-adapted) state by night and in the light-adapted state by day, even when the animal is kept in constant light or darkness.

It is thus possible by suitable preliminary treatment to arrange to have animals in which the pigment dispersion is in a variety of different states. Animals fully adapted to darkness will have both sets of pigments in a dark-adapted state. Those first dark-adapted and then placed for a few minutes into light will have the proximal pigments in the light-adapted state and the distal pigment in the dark-adapted state. Finally, animals kept in the light will have both pigment cells in the light-adapted state by day but have the distal pigment in the dark-adapted state during the time when it is night outside. In some ingenious experiments de Bruin and Crisp[125] have made use of these various situations to investigate further the function of the two groups of pigment cells. Prawns were placed in the middle of a glass dish around which was rotated a cylinder marked with alternate black and white stripes. In such a situation a prawn will either display nystatic (i.e. following movements and then a flick back to the previous position) movements of the eyestalk or will swim so as to keep pace with the moving cylinder *provided it can see the stripes*. A measure of eye function is thus obtained by studying the behaviour at a series of different light intensities with the stripes of various thicknesses and widths of separation. The results of the experiments indicated (1) that the important factor in light adaptation is movement of the proximal pigment, (2) that the visual acuity is lower in dark-adapted animals than in those which are light-adapted (3) that the eye is most sensitive to light when in the dark-adapted state; only dark-adapted animals responded when experiments were carried out in a dim light. The role of the white pigment is clearly that of a reflector, and it is probable that the loss of visual acuity in dark-adapted animals is due to the back reflection of some of the incident light from this pigment layer into neighbouring ommatidia.[125]

In neither the apposition nor the superposition eye is there any means of altering the focal distance of the lens system.   The position of the focused image will therefore vary according to the distance of each object under examination from the surface of the eye.   This is probably immaterial;   because the length of the rhabdomes is adequate to allow for a considerable degree of variation in image position.

Reception of the Image and Initiation of the Sensory Response

The crustacean eye contains considerable quantities of vitamin A;   and a rhodopsin-retinine system comparable with that responsible for the initial photo-chemical response of vertebrate rods has been located in *Homarus*.[529]   The main concentration of rhodopsin in this animal appears to be in the rhabdome, making it likely that the initial response to light occurs in the components of this structure, the rhabdomeres.   In vertebrates, when a molecule of rhodopsin absorbs a quantum of light energy, a series of reactions take place, resulting ultimately in the production of retinine and opsin[528]:

$$\text{Rhodopsin} + \text{light energy} \xrightarrow{\text{isomerisation}} \text{lumino-rhodopsin}$$
$$\rightarrow \text{metarhodopsin} \rightarrow \text{retinine} + \text{opsin}.$$

In *Homarus* too, the end products of exposure of rhodopsin to light are retinine and opsin, whilst metarhodopsin is an intermediate.[529]   The manner by which this breakdown of the visual pigment initiates further response in the arthropod eye is not clear, the next known step in the excitation process being a sudden change in the membrane potential of the stimulated retinula cell.[75] In insects, the resting potential of the visual cells is comparable with that of muscle cells, being of the order of 50–70 mV (the inside of the membrane being negative to the outside).   On stimulation by light, there is a change in this resting potential proportional to the intensity of the light and the membrane is completely depolarized at high light intensities.[75]   If light is allowed to stimulate the cell for more than a brief flash, the action potential rises to a peak and then declines fairly quickly to a maintained level of depolarization (fig. 63A).[75]   The time lag

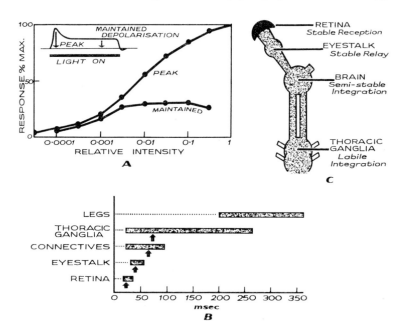

FIG. 63.   Electrical responses following light stimulation of the eye. A. The relationship between light intensity and size of visual cell action potentials.   B. Diagram of the relative latencies for responses at different levels of the central nervous system.   Arrows indicate average latencies.   C. Diagram showing the functional roles of various parts of the central nervous system in mediating visual responses. (A. modified from Burkhardt,[75] B. and C. from Camougis.[83])

between illumination of the eye and the origin of the electrical changes is quite short, of the order of 22 msec. in the crayfish. From the retina impulses are passed into the integrative system of the eyestalk and, if of sufficient intensity, will initiate axonal activity, which can be picked up also in other parts of the central nervous system (fig. 63B).[83]

Spectral Sensitivity and Colour Vision

The absorption spectrum for crustacean rhodopsin has a peak at about 500 mμ and also in the ultra violet; absorption falls off rapidly towards the red end of the scale (fig. 64).[529]   In agreement with

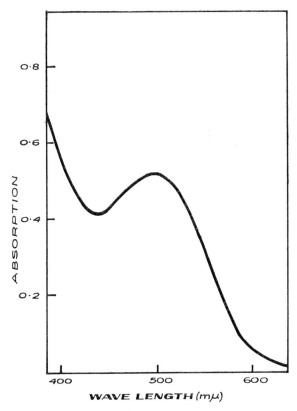

Fig. 64.   The absorption spectrum of a fresh preparation of lobster rhodopsin at pH 9·3 and 29°C.   (Simplified from Wald and Hubbard.[529])

this the behavioural response of many phototactic crustaceans indicates that they are most sensitive to wavelengths in the green and green/yellow.   Nevertheless certain species, such as *Artemia*, can clearly detect the presence of light of quite long wavelengths and will display a dorsal light-reflex in red light provided that the eyes are intact.[335] One possible reason for the greater relative sensitivity of the eye than that predicted from the response of rhodopsin to equal quanta of different wavelengths is that there will be a tendency for light of long wavelengths (e.g. red light) to pass through the pigment cells and hence stimulate the sense cells of neighbouring ommatidia, whilst short wavelengths cannot so pass.[75]

Even though it can be shown that animals respond to light over a wide range of wavelength, this does not prove that they are able to recognize colour differences. To demonstrate this involves testing the ability to discriminate between two coloured objects under conditions where the intensity of light reflected from both is the same. Critical studies indicate that at least copepods and cladocerans in addition to decapods can detect colour differences.

In the fly *Calliphora* three types of receptor cell have been found which have maximum sensitivities at different wavelengths (470, 486 and 521 mμ respectively). Of these the green receptors (486 mμ) are by far the most numerous and are the only ones in the upper part of the eye. In the rest of the eye, the evidence suggests that in each ommatidium there are five green cells, one blue and one yellow/green.[75] No comparable studies are yet available for the individual sense cells in the crustacean ommatidium.

Polarized Light Reception

Light emitted by the sun vibrates in all planes, but indirect light received on earth from scattered light in the sky is plane polarized, i.e. it vibrates in a single plane. The angle of vibration to any chosen axis is a function of the position of the sun to the part of the sky from which the polarized light is received. If the plane of polarization of light from different regions of the sky can be analysed, the position of the sun can therefore be determined even when obscured by cloud. In his classical work on bees von Frisch has shown that the position of the sun is used in navigation and that polarized light may be utilized when the sun is obscured. A number of Crustacea, both aquatic (*Daphnia*, mysids etc.) and terrestrial (Tallitrids), are known to use also a

sun (or moon) compass and polarized light in orientating themselves.[398, 399]   In the case of the Tallitrids, the animals usually
lie up during the day under cover in moist micro-habitats near the
strand line.   At night they move inland for 10 to 100 metres to
feed, returning before daybreak along a line perpendicular to the
shore.   There is good evidence that for this return migration
the animals make use of a moon compass,[398] moving at an angle to
the moon which is constantly adjusted to take account of the moon's
apparent passage across the sky.   They can be made to change
course if the true direction of the moon is concealed and its light is
reflected onto them from a mirror.   In the absence of moonlight
the animals are disorientated.   Following moonless nights, it
is essential that the animals find an inland moist micro-habitat
or return to the shore before the heat of the day.   It may be
presumed, therefore, that it is for this latter reason that they are
also adapted to make use of a sun compass and polarized light as
directional aids.   Proof that polarized light can be used is given
by the fact that the animals will still make the appropriate orientation when they are in shade, but that this orientation is disturbed
when the light reaches them only through polaroid filters.

The question whether the whole eye or only a single ommatidium is necessary for analysis of the plane of polarization has not
been satisfactorily answered for Crustacea; but in *Calliphora* the
size of the action potential produced by a single cell has been found
to vary with the direction of polarization in the case of a blue or
white light, suggesting that it is at the level of the single cell that
the analysis is based.[75]   The manner in which this analysis is
achieved is unclear.

Summing up the function of the insect ommatidium, Burkhardt[75] has suggested that the five green-type receptors deliver a
pattern of excitation to the central nervous system thus signalling
the direction of polarized light by means of their differing excitation.   Comparison of the overall level of excitation of the five
green type, the blue, and the green/yellow, determines the wavelength of the incident light; and the sum of the responses of all
seven cells indicates the intensity of the light.   Insufficient
evidence is available to determine whether or not such an inter-
pretive system could operate also in the crustacean eye, though
this would appear to be inherently reasonable.

HEARING

Quite a number of Crustacea can emit considerable volumes of sound as a result of stridulation (*Coenobita*, *Ocypode* and *Panulirus*) or raptorial clicks (the mantis shrimp, *Gonodactylis oerstedii* and the pistol shrimps *Alpheus* and *Synalpheus*). The latter produce such a barrage of sound in the frequency range from zero to 52 kc that it was not unknown during the war for inexperienced hydrophone operators to imagine that they were listening to an enemy secret weapon. A grumbling sound of unknown origin is produced by the lobster *Homarus* and can be detected readily by touching the carapace with a rod, the other end of which is held to the ear.[69]

It is curious, in view of the quantity and quality of sound produced by crustaceans, that although they have been shown to possess the ability to perceive low frequency vibrations transmitted via the substrate,[105] there is little evidence to suggest that they can detect air- or water-borne sound waves.

CHEMO-RECEPTION

Crustacea have long been known to be able to detect the presence of food at a distance, thus indicating the possession of a chemo-receptive sense.

Behavioural experiments suggest that the sense organs responsible for chemo-reception are more widely distributed on the body than those mediating the corresponding smell/taste sense in ourselves. Apart from the mouth region, the outer flagellum of the antennule, the chelipeds and dactylus of the walking legs seem to be well equipped with sensory endings able to initiate behavioural responses to chemical substances. For example, when glutamic acid is applied to the chelipeds of *Carcinus*, they are moved towards the mouth.[328]

Recordings taken of the activity of the sensory axons leading from the dactylus of the walking legs in *Carcinus*, *Portunus* and *Homarus* indicate that the sense organs are most sensitive to the presence of trimethylamine oxide (T.M.O.) and betaine and also, but to a lesser degree, to glutamic acid. The concentrations required to initiate responses are quite high (0·1–0·01 molar) suggesting that this is a contact taste receptor rather than a distance receptor.[328]   The whole study of chemo-receptors in arthropods is bedevilled by interference due to the fact that disturbance resulting from the introduction of the test solution tends to stimulate mechano-receptors.   In the case of the dactylus receptors, however, the interference is minimal, since there is a latent period of about 10–15 seconds after the application before the chemo-sensory response begins.   After this delay the chemo-sensory neurones' firing rate then gradually increases and reaches a maximum after about 50 seconds.   Adaptation of the sense organ is slow as long as the stimulating substance is present.

The limb chemical sense seems to be fairly specific since cystine, gamma-amino-butyric acid, aspartic acid and glycine produce no response.   The restriction to TMO and betaine is interesting and probably functionally important, since these substances are present in quite high concentrations in likely food organisms (TMO in fish and crustacea, betaine in many invertebrates).[328]

The precise location of chemo-sensory cells on the limbs in crustacea is as yet uncertain, but small pore canals which penetrate the cuticle and contain a single neurone may be implicated.   In insects, it is known that the chemical sense is associated with the base of special hairs.   Typically, the sensory hairs in an insect such as the blow-fly have five neurones at the base and the dendrites of these cells extend up the hair.   Recordings taken from the axons of these neurones indicate that each is specialized for one particular sensory function (though gross stimulation can activate more than one receptor).   Three of the cells are known to be taste receptors responsible respectively for the detection of sugar, salt and water, whilst a fourth neurone is a mechano-receptor.   The function of the fifth neurone, where present, is still uncertain.[138]   Although work on crustacean chemo-sense has not reached such an advanced state as on that of insects, the behavioural evidence suggests that the limbs of crustaceans are also capable of detecting salts and water.   Thus, the littoral isopod *Ligia* can differentiate between salt solutions and distilled water and between NaCl and $CaCl_2$ when allowed to walk across filter paper soaked in the test substance.[24, 25]   Terrestrial isopods too have a well developed capacity to recognize the degree of humidity of their environment, since they tend to congregate in the moister parts of a humidity chamber. Even aquatic forms appear to possess a water receptor; for it has been observed that if *Asellus aquaticus* are placed on a moist spot on filter paper they may walk around within the confines of the damp area and will generally not leave it until the spot can no longer be seen by the human eye to be wet. Should they stray over the boundary onto the dry filter paper, they will stop, pause, and return into the moist area.[336] It would be interesting to know more about the chemosensory basis of the mechanisms which enable some estuarine forms, such as *Gammarus* spp., to select particular concentrations when they are placed in a salinity gradient.[317] The location of salinity-sensing systems in estuarine forms has not been studied, but it has been found that such a sense is present in the antennulae of the marine crabs *Podophthalmus* and *Portunus*. In these crabs a nervous response is elicited when small droplets of diluted sea water are placed on the flagellum of the antennule, and the type of response is related to the salinity used. No response occurs when the drops are placed on the basal segments of the appendage.[156b]

One of the more interesting chemo-sensory specializations is that of *Platyarthrus hoffmansegii*, which is attracted by formic acid but not by other organic acids. This is no curiosity, however, but rather an adaptional feature since this isopod is an inquiline of ants' nests.

On the outer flagellum of the antennule of decapods there are specialized groups of hairs, the aesthetascs, which on morphological grounds have long been suspected of chemo-sensory function (fig. 65). This has now received physiological support.[330] These hairs are arranged in rows and, unlike other crustacean hairs, are thin-walled and unpigmented. They are restricted to the outer flagellum, which is chemo-sensory, being absent from the inner flagellum which is not. Each hair has a large number, perhaps 120–150, of neurones at its base, so that a hair row will contain some 4,000 neurones, and the flagellum as a whole about half a million.

Mechanical stimulation of the hairs does not excite the neurones but they become active in the presence of TMO, betaine (and macerated anchovies!). Such a large accumulation of neurones presumably represents a considerable potential for the detection of numerous compounds.[330]

On either side of the hair tufts on the antennule are guard hairs which are primarily mechano-detectors, and also companion hairs. The latter may have four neurones at their base and, since never more than two cells are involved in mechano-reception, it is

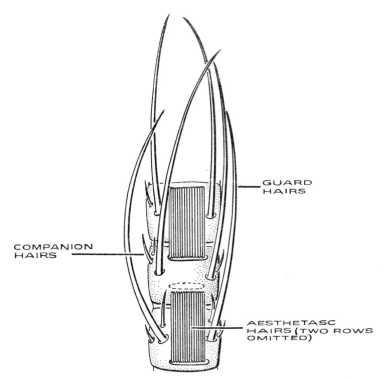

GUARD HAIRS

COMPANION HAIRS

AESTHETASC HAIRS (TWO ROWS OMITTED)

FIG. 65. The antennular chemosensory organs of *Panulirus*. Diagram of the ventral surface of the hair tuft region; the position of the two rows of aesthetasc hairs, omitted for clarity, is shown by a broken line. (From Laverack.[330])

possible that the others are involved in chemo-, thermo-, or osmo-reception.[330] Similarly the peg receptors on the chelae also have 3–5 neurones not all of which are mechano-receptors.[326]

So far, only a start has been made to the study of crustacean chemo-sensory systems, and much more remains to be discovered. For example, nothing is known of the manner in which the sense organ itself is excited by stimulatory material.

9 : Feeding and Digestion

The food and feeding habits of Crustacea are as varied as their ways of life. Many of the decapods are omnivorous. This must not be taken as necessarily indicating that all is grist to the gastric mill, however, for while some, such as *Carcinus, Nephrops* and *Homarus*, are scavengers, favouring a mainly meat diet, others, such as *Maia*, have a greater predilection for vegetation. True vegetarians are rare among the higher Crustacea. Wood lice, it is true, will sometimes attack living plants; but both they and the terrestrial amphipods feed primarily on rotting vegetation. Individual choice is sometimes important even where there is apparently parity of opportunity to select the diet. For example, in a population of *Gammarus duebeni* studied by Hynes,[274] some were found to have eaten only animal material, some only plant material, and some both.

The marine isopod *Limnoria* (the gribble) finds both food and shelter by tunnelling into wood piles and, where numerous, can cause severe damage to wooden harbour installations.

The herbivores *par excellence* of the Crustacea are found amongst the Cladocera and Copepoda, whose diet consists largely of unicellular algae, though many are not averse to supplementing this with planktonic animals and even smaller members of their own species. Cannibalism, in fact, has been noted in many crustacean groups. It is well known that crabs will attack and eat any individuals which are incapacitated by moulting or damage; but many other forms also, such as *Asellus aquaticus* (Isopoda), *Neomysis integer* (Mysidacea), *Chiridius* (Copepoda), will attack their fellows, especially if other forms of food are lacking. Crabs are often partial to decaying flesh; *Asellus* and *Neomysis* are more fastidious, eating their fellows only when alive or recently dead.

1. Method of Feeding

Decapods, such as *Nephrops*, seize their food with their chelae and pass it directly, or via the small chelae on the fourth walking legs, to the third maxillipeds. These grip and push it towards the mandibles which slice and grind it prior to swallowing. The size of particles swallowed varies with the group. The mandibles of the Reptantia merely slice the food into chunks, which are swallowed and chewed into smaller particles by the chitinous teeth of the gastric mill. By contrast, the mandibles of the Natantia themselves masticate the food into fine particles, and the gastric mill of these forms is thus rendered unnecessary and is reduced.

FILTER-FEEDING

The filtering of particles from the water is practised mainly by the smaller species, Ostracoda, Copepoda, Cladocera, etc. and sessile forms such as the Cirripedia; but even in the Malacostraca some species, e.g. *Callianassa*, *Upogebia* and mysids, feed either wholly or in part by this means. The fact that particles are filtered from a stream of water must not be taken to indicate that the feeders are totally unselective in their diet except on the basis of particle size. Some filter feeders certainly do display the capacity to take one type of particle and reject others. For instance *Calanus* takes the alga *Lauderia* in preference to *Chaetoceros*.[243] *Evadne* is another selective feeder, taking *Ceratium furca* and *Peridinium* but ignoring diatoms and even other species of *Ceratium*.[19]

The mode of filter feeding is variable; barnacles, such as *Balanus*, rhythmically sweep the water with their long hair-fringed thoracic appendages, withdrawing them at the end of each sweep, to remove trapped particles. Such vigorous searching of the water for food is necessary in forms which can exist so far up the beach that they may be covered by water for only a few hours of the day; but it would clearly be wasteful of energy in continuously

submerged pelagic forms. Indeed a specimen of *Concoderma**
showed a more leisurely mode of feeding; the appendages were
spread as a net and only periodically withdrawn. Occasionally,
however, a single appendage flicked at high speed towards the
mouth, presumably after being touched by particles in the water.

The trunk limbs of anostracan branchiopods, such as *Artemia*,
are used to create a current of water in the gully between the
bases of the legs. Fine setae on the appendages prevent the
escape of any particles contained in this current and the food is
ultimately gathered together and taken to the mouth along a
groove on the ventral surface. Modifications and specializations
of this general technique are found in *Daphnia* and other cladocerans. Here the third and fourth trunk limbs combine the functions of producing the current and straining off the particles of
food, whilst bristles on the first and second limbs prevent exces-
sively large particles from entering the system.

By contrast with the branchiopods, copepods produce the
feeding current by movement of the antennae, mandibles and
maxillae, and filter off particles with the maxillae. Mysids also
use the setae on the maxillae for straining off particles but in the
case of these forms the current is produced by the exopodites of
the thoracic legs. Filter feeding in mysids is frequently supplemented by the direct seizure of larger particles by the mouthparts. In mature females of *Neomysis* the latter method pre-
dominates, possibly because the brood pouch interferes with the
direction of the filter feeding current.[99] A comparable effect has
been observed in the larger euphausids such as *Meganyctiphanes*,
where the smaller individuals are filter feeders and the larger
ones primarily carnivorous.[366]

The mud-shrimp *Upogebia pugettensis*, which constructs
burrows some feet long, is interesting in that its first two pairs of
limbs are specialized both for carrying mud and for filter feeding,
though both functions, of course, cannot be carried on simultaneously. When digging, the limbs act as a mud basket for removal
of the spoil; at other times they filter particles from the constant

---

* It was obtained by the kind permission of Col. Hasler from the
bottom of the yacht Jester on his return from the single-handed Atlantic
Ocean race.

stream of water (created by other limbs) which is caused to pass through the burrow.

2. Structure of the Gut

The gut of Crustacea is usually a straight tube, though in some cladocerans it may be coiled.   It is divided into three main regions, the fore-gut (stomodaeum), mid-gut (mesenteron) and hind-gut (proctodaeum).   The fore- and hind-guts arise from ectodermal intuckings, and their linings secrete cuticle.   The mid-gut lining is mesodermal and lacks cuticle.

In the decapods the fore-gut is sub-divided into an oesophagus, cardiac-stomach and pyloric stomach (fig. 66).   Stomachs are also present in many other forms, though in some cases the mesenteron as well as the fore-gut may contribute.   The structures are therefore not homologous with those of the decapods. The cardiac stomach of decapods is large and sack-shaped.   It has a thick chitinous layer, which may be calcified in certain regions to form ossicles which are the masticatory apparatus of the gastric mill.   Muscles move the ossicles to grind the food into fine particles before it is passed to the pyloric stomach.

The gastric mill tends to be especially well developed in those species which swallow large chunks of food, e.g. most Reptantia. Apart from the Malacostraca, only in the Ostracoda is it apparent.

Food leaving the cardiac stomach is passed through the cardio-pyloric valve which permits the passage of only small particles.

Behind the pyloric stomach is the mid-gut and its lateral expansions, the hepatopancreas or mid-gut gland.   The mid-gut itself is of variable length, being long in *Homarus* but short in crabs.   The hepatopancreas is a mass of fine, blindly ending tubules which unite into a common hepatopancreatic duct opening into the anterior end of the mid-gut.   A complex chitinous filter formed by the fore-gut protects the entrance of channels leading into the hepatopancreatic duct, and this prevents all but the finest particles from passing up the duct to the absorptive cells of the hepatopancreatic gland (fig. 66*B*).

In *Nephrops* an anterior dorsal caecum and a posterior diverticulum, which open at the junction of the mid- and hind-guts,

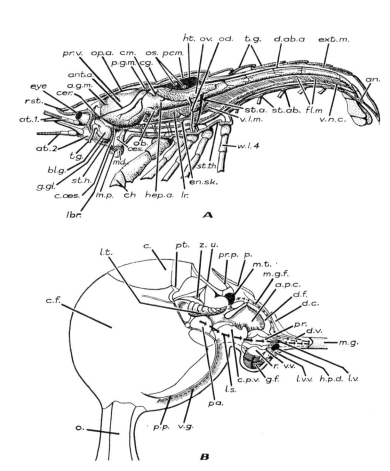

FIG. 66. *A*. Side view of the crayfish to show a generalized view of the gut. *B*. A detailed diagram of the foregut of *Nephrops*: *a.p.c.* anterior pyloric chamber; *c.* cardiac ossicle; *c.f.* cardiac foregut; *c.p.v.* cardio-pyloric valve; *d.c.* dorsal caecum; *d.f.* dorsal fold; *d.v.* dorsal valve; *g.f.* gland filter; *h.p.d.* hepatopancreatic duct; *l.* labrum; *l.s.* lateral food stream; *l.t.* lateral tooth; *l.v.* lateral valve; *l.v.v.* lower ventral valve; *m.g.* mid gut; *m.g.f.* mid gut filter; *m.t.* median tooth; *o.* oesophagus; *p.* pyloric ossicle; *p.p.* post-pectineal ossicle; *pa.* pads; *pr.* press; *pr.p.* prepyloric ossicle; *pt.* pterocardiac ossicle; *r.* ridge of gland filter; *u.* urocardiac process; *v.g.* ventral groove; *v.v.* ventral valve; *z.* zygocardiac ossicle. (From Yonge.[564])

258

apparently provide additional absorptive surface.[564]   The hind-gut is a short, stout tube opening at the anus.

Cilia are of course absent in the Crustacea, and so all movement of food in the gut is by peristalsis and anti-peristalsis. Peristaltic and anti-peristaltic waves are apparent in the mid-gut, and are even more pronounced in the hind-gut, where in *Nephrops* they may occur from three to five times a minute.[564] Presumably such co-ordinated muscle movements are under the control of the extensive nerve plexus in the gut wall.

A curious feature of the muscles responsible for peristalsis is that whilst in the mid-gut the longitudinal muscles lie outside the circular muscles, the reverse arrangement is present in the hind-gut.[564]   Unlike the situation in vertebrates, the gut-muscles are all striated.

3. Digestion and Absorption

The evidence at present available suggests that the digestive enzymes are produced mainly, possibly even solely, by the hepato-pancreas.   The secretion of this gland, however, can be passed forward into the cardiac stomach and initiate digestion in this region of the fore-gut.   The pH of the fluid in the cardiac stomach of *Nephrops*, is slightly acid[564] which is suitable for the pH optimum of the tryptic proteolytic enzyme present.[564]   This enzyme will digest proteins such as fibrin, casein and peptone.   The comparable enzyme responsible for caseinase activity in the crayfish *Orconectes virilis* is found to have two pH maxima at pH 6 and pH 7·5–8·5.[139]   There does not appear to be any enzyme comparable with vertebrate chymotrypsin in *Orconectes*, since the specific substrate for chymotrypsin, BTEE (N-benzoyle tyrosine ethyl ester), is not hydrolysed by the gastric extract.[139]   Pepsin-like enzymes also appear to be lacking.[526]

Abundant enzymes are commonly present, breaking down carbohydrates and related compounds.   Thus the stomach fluid of *Astacus* contains cellobiase (cellobiose→glucose), alpha-galactosidase (melibiose→galactose+glucose), saccharase (invertase) (sucrose→glucose+fructose), beta-galactosidase (lactose→galactose+glucose), maltase (maltose→glucose), isomaltase (dextrans and isomaltose→glucose) alpha-xylosidase (xylans→xylose), amylase (starch→maltose), beta-1, 3-glucanase (glucan→glucose), beta-1, 4-glucanase (cellulase) (cellulose→isomaltose), mannanase (mannan→mannose), xylanase (xylan→xylose) and a chitinase.[304] Most of these enzymes, including cellulase and chitinase, are also present in extracts of the hepatopancreas. It is probable, therefore, that the enzymic activity is not due to bacterial activity. Naturally this is difficult to prove categorically, but at least seems fairly certain in the case of the chitinase, as bacteria plated from the gut do not show the ability to attack chitin. By contrast, the chitinase of *Porcellio scaber* does seem to be produced by bacteria.[276] The pH optima of the cellulase and chitinase are lower than that of the proteases already mentioned, being 4·0–4·5 and 3·0–4·0 respectively.[304]

A cellulase is apparently absent in *Nephrops*[563] and *Ligia oceanica*[385] but, as might have been expected, occurs in extracts of the hepatopancreas (digestive tubules) of the wood-boring isopod *Limnoria lignorum*.[430] Here again, no bacteria showing cellulase activity can be found in the gut[430]; so the enzyme is almost certainly produced by the gland itself.

Starch, glycogen, sucrose, maltose and lactose, but not inulin or raffinose, are digested by *Nephrops* hepatopancreas extracts.[564] Trehalase occurs in the guts of the woodlice *Porcellio* and *Oniscus*. Alginic acid and pectin are not digested by *Astacus*[304]; but there is no evidence to show whether more fully vegetarian forms such as *Maia* have the necessary enzymes or not.

In *Nephrops* there is a lipase which will split olive oil; various esters, such as methyl acetate, amyl acetate, ethyl acetate and butyl acetate, are also digested. Butyrin from milk is converted to butyric acid.[564] Young[564] failed to demonstrate the presence of bile salts in *Nephrops*, but their presence in crayfish has since been demonstrated by Vonk.[525, 527]

ABSORPTION

The fore- and hind-guts of Crustacea are reported to act merely as semi-permeable membranes,[564] and there is as yet no evidence that they can absorb material other than water. Thus, when glucose is introduced into the fore-gut and this region of the gut is ligatured, the glucose is not absorbed. Similarly olive oil and ferrous lactate are not taken up by either the fore- or hind-guts.[564] This is in contrast with the insect gut (cf. Treherne[512]), where numerous materials can be absorbed from the hind-gut.

The mid-gut and hepatopancreas are therefore the primary absorptive areas. Digestive products and fine particles are passed

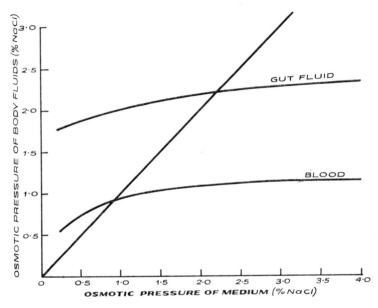

FIG. 67. The relation between the osmotic pressure of the body fluids and the medium, in *Artemia*. (From Croghan [117])

up the hepatopancreatic duct into the gland. Digestion is almost wholly extra-cellular in arthropods, unlike the case in the majority of invertebrates. This has complicated the identification of the cells actually responsible for absorption. In consequence, there has been some controversy as to whether cells in the hepatopancreas combine secretory and absorptive functions, or are specialized for either one function or the other. As frequently happens in biological controversies, it has now been found that there was a measure of truth on both sides. Thus, the most recent study of the problem[122] has shown that all cells in the hepatopancreas arise from embryonic cells at the end of the tubules and that each cell is at first absorptive, then secretory, then fibrillar. The production rate of cells is greater than the death rate, so that tubules elongate as the animal increases in size.

Oil droplets are absorbed in the mid-gut.[564]

Salts and water are actively taken up from the gut by forms whose blood can be maintained hypotonic to the medium. Such forms drink the medium either orally or anally,[117, 204, 428] and, if the medium is dyed, the dye becomes concentrated in the gut as the salts and water are absorbed. Careful measurements of the osmotic pressure of the gut fluid in *Artemia*[115] indicate that, although it is less concentrated than the medium, it is considerably more concentrated than the blood (fig. 67). Clearly, therefore, the salts must be being taken up from the gut fluid. The manner in which water is taken up is more obscure. Either the gut wall is able to transport water actively against the concentration gradient, as is shown to be possible in the insect rectum,[413] or alternatively, the osmotic concentration immediately in contact with the gut cells may perhaps be lowered below the blood concentration, so that the water can be taken up passively by osmosis, even though the bulk of the fluid in the gut has a higher concentration than the blood.

EFFICIENCY OF DIGESTION

Studies on feeding rates and the efficiency of digestion have been confined almost solely to the copepods.

Many of the observations and calculations on the amount of water swept clear daily by a form such as *Calanus* are conflicting, ranging from about 5 cc to over 1,000 cc per day. A recent study,[111] however, indicates that the average amount of food taken in daily during the summer months by female *Calanus helgolandicus* is equivalent to some 25% of the dry body weight. Naturally, when such high rates of food intake are occurring, passage of food along the gut is very rapid. The gut of a starved copepod will be filled within one hour when offered food and will be emptied in one hour when food is withdrawn.[218] Digestion therefore has to be very efficient if all the material taken in is to be utilized in the short time available. In fact utilization is not

complete.   When feeding on an abundant source of phytoplankton, *Calanus* may produce faecal pellets at the rate of one every five
to seven minutes.[361]   These pellets are green and, perhaps not
surprisingly, sometimes contain undamaged cells.   It has even
been possible to start a new phytoplankton culture from viable
cells taken from the faeces.[242]   Nevertheless, the digestive processes are quite efficient, since some 50–90% of the food taken
in is assimilated.[361]   The precise level, however, tends to vary
with the food organism offered, flagellates such as *Dicrateria*,
Chromulina and *Chlorella* being less effectively digested than
other forms.

4. Nutrition

Both rate of growth and egg-laying capacity have been used as
indices to test the suitability of various diets; and the results
indicate that, at least in the case of some omnivores, not all requirements are met by a single food type.   For example,
Gammarus fed solely on animal food grow less rapidly than those
fed on both animal and plant materials (fig. 68).   Conversely,
larval *Palaemonetes* reared solely on plant material (*Nitzchia*,
Chlamydomonas, *Nanochloris*, *Porphyridium* and *Pyramimonas*)
do less well than those given animal food also, in the form of
Artemia nauplii.[51]

Starved copepods stop laying eggs.   Egg-laying capacity is
restored in the case of *Calanus* by feeding with diatoms such as
Chaetoceros, *Skeletonema* or *Rhizosolenia*; but flagellates such
as *Chlorella*, *Hemiselmis* and *Dicrateria* are apparently lacking in
some essential food factor, since if fed alone to *Calanus* they do not
restore egg-production in starved individuals.[360]   Indeed, *Chlorella*
appears to pass undigested through the gut.   Growth and reproductive capacity of the harpacticoid *Tigriopus japonicus* have
also been found to vary widely according to the species of algae
with which it is fed.[488]

Nutritional requirements vary at different times of the year.
Poikilotherms do not require more calories during winter months
to maintain the body temperature as homiotherms do; instead,
they are less active, and so have lower basic metabolic rates.
Growth and reproduction also tend to be slower at low temperatures, again economizing on food requirements during the winter.

Nevertheless, as was pointed out by Raymont,[431] there is still an apparent discrepancy between the maintenance requirements of copepods for food and the amount actually present in the water during winter. This discrepancy may be more apparent than

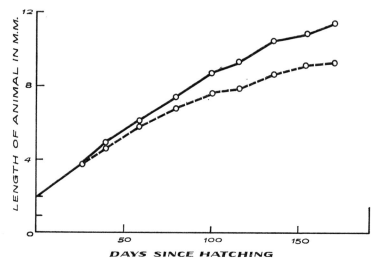

FIG. 68.   Retardation of growth of *Gammarus* caused by deficiency of plant food. - - - - - Animals fed animal materials alone, ——— Animals fed on plant and animal foodstuffs. (From Kinne.[294])

real; for over-wintering copepods show little reduction in the fat content (stage V *Calanus*) during the winter months; which implies that the food intake is adequate to supply the needs of the animal without recourse to reserves.

5. Feeding Migrations

Many species make migrations from one place to another in connection with either feeding or reproduction. In the latter category are the land crabs which migrate *en masse* to the sea to reproduce, on which occasions they have been described as sounding ' like a regiment of Cuirassiers on the march ', as they bounce from rock to rock on their headlong descent to the water. The Chinese wool-handed crab, *Eriocheir* makes an extensive feeding migration to fresh water, young crabs penetrating many tens or even hundreds of miles up the North European rivers. They return ultimately, however, to saline waters to breed.

The diurnal migration of many planktonic forms, especially copepods, is well known.[431]   Even such small forms as the copepod *Calanus* may migrate daily, from a depth of 400 metres during the day to near the surface at night, whilst migrations from 200 to 1,000 metres are made by many euphausids, decapods, mysids and amphipods.

Finally, we must mention the extraordinary migration of the American spiny lobster *Panulirus* though its precise purpose remains in doubt.   On these migrations the animals move along the bottom parallel with the shore in chains of 3–30 individuals, the members composing each chain positioning themselves in single file, each with its cephalothorax lying immediately above the tail of the one in front.[250]   This crustacean ' Conga ' lacks little but the rhythm.

References

1. ABRAMOWICZ, A. A., HISAW, F. L. and PAPANDREA, D. 1944. The occurrence of a diabetogenic factor in the eyestalks of crustaceans. *Biol. Bull. Woods Hole*, **86** : 1–5.

2. ALEXANDROWICZ, J. S. 1951. Muscle receptor organs in the abdomen of *Homarus vulgaris* and *Palinurus vulgaris*. *Q. Jl microsc. Sci.* **92** : 163–200.

3. ———— 1952. Receptor elements in the thoracic muscles of *Homarus vulgaris* and *Palinurus vulgaris*. *Q. Jl. microsc. Sci.* **93** : 315–346.

4. ———— 1952. Muscle receptor organs in the Paguridae. *J. mar. biol. Ass. U.K.* **31** : 277–286.

5. ———— 1953. Nervous organs in the pericardial cavity of the decapod crustacea. *J. mar. biol. Ass. U.K.* **31** : 363–380.

6. ———— 1954. Notes on the nervous system of Stomatopoda. IV, Muscle receptor organs. *Pubbl. Staz. zool. Napoli*, **25** : 94–111.

7. ———— 1956. Receptor elements in the muscles of *Leander serratus*. *J. mar. biol. Ass. U.K.* **35** : 129–144.

8. ———— 1958. Further observations on proprioceptors in crustacea and a hypothesis about their function. *J. mar. biol. Ass. U.K.* **37** : 379–396.

9. ALEXANDROWICZ, J. S. and CARLISLE, D. B. 1953. Some experiments on the function of the pericardial organs of crustacea. *J. mar. biol. Ass. U.K.* **32** : 175–192.

10. ALEXANDROWICZ, J. S. and WHITTEAR, M. 1957. Receptor elements in the coxal region of decapod crustacea. *J. mar. biol. Ass. U.K.* **36** : 603–628.

11. AMBERSON, W. R., MAYERSON, H. S. and SCOTT, W. J. 1924. The influence of oxygen tension upon metabolic rate in invertebrates. *J. gen. Physiol.* **7** : 171–176.

12. AMAR, R. 1948. Un organe endocrine chez *Idotea* (Crustacea isopoda). *C. R. hebd. Séanc. Acad. Sci. Paris*, **227** : 301–303.

13. ARMITAGE, K. B. 1962. Temperature and oxygen consumption of *Orchomonella chilensis* (Heller) (Amphipoda : Gammaroidea). *Biol. Bull. Woods Hole*, **123** : 225–232.

14. ARUDPRAGASAM, K. D. and NAYLOR, E. 1964. Gill ventilation and the role of reversed respiratory currents in *Carcinus maenas* (L). *J. exp. biol.* **41** : 299–308.

15. 1964. Gill ventilation volumes, oxygen consumption and respiratory rhythms in *Carcinus maenas* (L). *J. exp. Biol.* **41** : 309–322.

16. ARVEY, L., ECHALIER, G. and GABE, M. 1956. Organe Y et gonade chez *Carcinides maenas* L. *Annls Sci. nat. Zool. Biol. Animal.* **18** : 263–268.

17. ATTWOOD, H. L. 1965. Characteristics of fibres in the extensor muscle of a crab. *Comp. Biochem. Physiol.* **14** : 205–207.

17b. ATTWOOD, H. L. and HOYLE, G. 1965. A further study of the paradox phenomenon of crustacean muscle. *J. Physiol.* **181**: 225–234.

18. AYERS, J. C. 1938. Relationship of habitat to oxygen consumption in certain estuarine crabs. *Ecology*, **19** : 523–527.

19. BAINBRIDGE, V. 1958. Some observations on *Evadne nordmanni* Loven. *J. mar. biol. Ass. U.K.* **37** : 349–370.

20. BAINBRIDGE, R. and WATERMAN, T. H. 1957. Polarised light and the orientation of two marine crustaceans. *J. exp. Biol.* **34** : 342–364.

20b. BAKER, P. F. 1965. Phosphorus metabolism of intact (spider) crab nerve and its relation to the active transport of ions. *J. Physiol.* **180**: 383–423.

21. BARNES, H. 1963. Light, temperature and the breeding of *Balanus balanoides*. *J. mar. biol. Ass. U.K.* **43** : 717–727.

22. BARNES, H. and GONOR, J. J. 1958. Neurosecretory cells in the cirripede, *Pollicipes polymerus*. *J. mar. Res.* **17** : 81–102.

23. BARNES, T. C. 1935. Salt requirements and orientation of *Ligia* in Bermuda. III. *Biol. Bull. Woods Hole*, **66** : 259–268.

24. 1939. Experiments on *Ligia* in Bermuda. VI. Reactions to common cations. *Biol. Bull. Woods Hole*, **76** : 121–126.

25. 1940. Experiments on *Ligia* in Bermuda. VII, Further effects of sodium, ammonium and magnesium. *Biol. Bull. Woods Hole*, **78** : 35–41.

26. BAUMBERGER, J. P. and DILL, D. B. 1928. Study of glycogen and sugar content and the osmotic pressure of crabs during the moult cycle. *Physiol. Zool.* **1** : 545–549.

27. BEADLE, L. C. and BEADLE, S. F. 1949. Carbon dioxide narcosis. *Nature, Lond.* **164** : 235.

28. BEAMENT, J. W. L. 1961. The water relations of insect cuticle. *Biol. Rev.* **36** : 281–320.

29. BEECHEY, R. B. 1961. Some spectrophotometric studies on respiring particles isolated from the hepatopancreas of *Carcinus maenas*. *Comp. Biochem. Physiol.* **3** : 161–174.

30.   1961. The difference spectrum of respiring particles isolated from the hepatopancreas of *Carcinus maenas*. *Biochem. J.* **78** : 3P.

31. BELMARICH, F. A. 1963. Biologically active peptides from the pericardial glands of the crab, *Cancer borealis*. *Biol. Bull. Woods Hole*, **105** : 367.

32. BERTALANFFY, L. VON, 1957. Quantitative laws in metabolism and growth. *Q. Rev. Biol.* **32** : 217–231.

33. BETHE, A. 1897. Das Nervensystem von *Carcinus maenas*. *Arch. microsk. Anat. EntwMech.* I, **50** : 462–546. II, **50** : 589–639. III, **51** : 382–452.

34. BIANCHI, C. P. and SHANES, A. M. 1959. Calcium influx in skeletal muscle at rest, during activity and during potassium ion contracture. *J. gen. Physiol.* **42** : 803–815.

35. BIELAWSKI, J. 1964. Chloride transport and water intake into isolated gills of crayfish. *Comp. Biochem. Physiol.* **13** : 423.

36. BLISS, D. E. 1951. Metabolic effects of sinus gland or eyestalk removal in the land-crab, *Gecarcinus lateralis*. *Anat. Rec.* **111** : 502.

37.   1953. Endocrine control of metabolism in the land crab, *Gecarcinus lateralis*. I, Differences in the respiratory metabolism of sinus glandless and eyestalkless crabs. *Biol. Bull. Woods Hole*, **104** : 275–296.

38.   1956. Neurosecretion and the control of growth in a decapod crustacean. In : *Bertil Hanstrom : zoological papers in honour of his sixty-fifth birthday*, ed. Wingstrand, K. G., Zoological Institute, Lund, Sweden.

39.   1959. In : *Physiology of insect development.*, ed. Campbell, Univ. Chicago Press.

40.   1960. Locomotor activity of land crabs during the premoult period. *Science*, **132** : 145–147.

41.   1963. The pericardial sacs of terrestrial Brachyura. Phylogeny and evolution of Crustacea. *Bull. Mus. comp. Zool. Harv.* (Special publication).

42. BLISS, D. E. and BOYER, J. P. 1964. Environmental regulation of growth in the decapod crustacean, *Gecarcinus lateralis*. *Gen. & compar. Endocr.* **4** : 15–41.

43. BLISS, D. E. and WELSH, J. H. 1952. The neurosecretory system of brachyuran Crustacea. *Biol. Bull. Woods Hole*, **103** : 157–169.

44. BOARDMAN, D. L. and COLLIER, H. O. J. 1946. The effect of magnesium deficiency on neuromuscular transmission in the shore crab, *Carcinus maenas*. *J. Physiol.* **104** : 377–383.

45. BOISTEL, J. and FATT, P. 1958. Membrane permeability change during inhibitory transmitter action in crustacean muscle. *J. Physiol.* **144** : 176–191.

46. BOWLER, K. 1963. A study of the factors involved in acclimatization to temperature and death at high temperature in *Astacus pallipes*. II, Experiments at the tissue level. *J. cell. comp. Physiol.* **62** : 133–146.

47. BOYD, C. M. and JOHNSON, M. W. 1963. Variations in the larval stages of a decapod crustacean, *Pleuroncodes planipes*, Simpson (Galatheidae). *Biol. Bull. Woods Hole*, **124** : 141–152.

48. BOYLAND, E. 1928. Chemical changes in muscle. Part II, Invertebrate muscle. Part III, Vertebrate cardiac muscle. *Biochem. J.* **22** : 362–380.

49. BRICTEUX-GREGOIRE, S., DUCHATEAU-BOSSON, GH., JEUNIAUX, C. and FLORKIN, M. 1962. Constituants osmotiquement actifs des muscles du crabe chinois, *Eriocheir sinensis*, adapté à l'eau douce ou à l'eau de mer. *Archs. int. Physiol. Biochim.* **70** : 273–286.

50. BRICTEUX-GREGOIRE, S., JEUNIAUX, C. and FLORKIN, M. 1961. Rôle de la variation de la composante amino-acide intracellulaire dans l'euryhalinité de *Leander serratus* L. et de *Leander squilla* L. *Archs. int. Physiol.* **69** : 744–745.

51. BROAD, A. C. 1957. The relationship between diet and larval development of *Palaemonetes*. *Biol. Bull. Woods Hole*, **112** : 162–170.

52. BRÖCKER, H. 1935. Untersuchungen über das Sehvermogen des Einsiedler-krebse, *Eupagurus bernhardus*. *Zool. Jb.* Abt. *Allgem. Zool. Physiol. der Tiere*, **55** : 399–429.

53. BROWN, F. A. 1934. The chemical nature of the pigments and the transformation responsible for color changes in *Palaemonetes*. *Biol. Bull. Woods Hole*, **67** : 365–380.

54. BROWN, F. A. and JONES, G. M. 1947. Hormonal inhibition of ovarian growth in the crayfish, *Cambarus*. *Anat. Rec.* **99** : 657.

55. ———— 1949. Ovarian inhibition by a sinus-gland principle in the fiddler crab. *Biol. Bull. Woods Hole*, **96** : 228–232.

56. BROWN, F. A. and MEGLITSCH, A. 1940. Comparison of the chromatophoretic activity of insect corpora cardiaca with that of crustacean sinus glands. *Biol. Bull. Woods Hole*, **79** : 409.

57. BROWN, F. A. and SANDEEN, M. I. 1948. Responses of the chromatophores of the fiddler crab, *Uca*, to light and temperature. *Physiol. Zoöl.* **21** ; 361–371.

58. BROWN, F. A., BENNETT, M. F. and WEBB, H. M. 1954. Persistent daily and tidal rhythms of oxygen consumption in fiddler crabs. *J. cell. comp. Physiol.* **44** : 477–506.

59. BROWN, F. A., SANDEEN, M. I. and WEBB, H. M. 1948. The influence of illumination on the chromatophore system of *Palaemonetes vulgaris*. *Anat. Rec.* **101** : 733.

60. BROWN, F. A., WEBB, H. M. and SANDEEN, M. I. 1952. The action of hormones regulating the red chromatophores of *Palaemonetes*. *J. exp. Zool.* **120** : 391.

61. BROWN, F. A. and WEBB, H. M. 1948. Temperature relations of an endogenous daily rhythmicity in the fiddler crab, *Uca*. *Physiol. Zoöl.* **21** : 371.

62. BROWN, F. A., WEBB, H. M. and BENNETT, M. F. 1955. Proof for an endogenous component in persistent solar and lunar rhythmicity in organisms. *Proc. nat. Acad. Sci. Wash.* **41** : 93.

63. BROWN, F. A. JR., HINES, M. N. and FINGERMAN, M. 1952. Hormonal regulation of the distal retinal pigment of *Palaemonetes*. *Biol. Bull. Woods Hole*, **102** : 212–225.

64. BRUNOW, H. 1911. Der Hungerstoffwechsel des flusskrebses (*Astacus fluviatilis*). *Z. allg. Physiol.* **12** : 215–276.

65. BRYAN, G. W. 1960. Sodium regulation in the crayfish, *Astacus fluviatilis*. III, Experiments with NaCl-loaded animals. *J. exp. Biol.* **37** : 113–128.

66. ———— 1960. Sodium regulation in the crayfish, *Astacus fluviatilis*, II, Experiments with sodium-depleted animals. *J. exp. Biol.* **37** : 100–112.

67. ———— 1960. Sodium regulation in the crayfish, *Astacus fluviatilis*. I, The normal animal. *J. exp. Biol.* **37** : 83–99.

67b. BUCHANAN, J. B. 1963. The biology of *Calocaris macandreae* (Crustacea: Thalassinidea). *J. mar. biol. Assoc.* **43** : 729–747.

68. BULLOCK, T. H. 1955. Compensation for temperature in the metabolism and activity of poikilotherms. *Biol. Rev.* **30** : 311–342.

69. ———— quoted by Moulton (1963). In : Acoustic orientation of marine fishes and invertebrates. *Erg. Biol.* **26** : 27–39.

70. BURBANCK, W. D., EDWARDS, J. P. and BURBANCK, M. P. 1948. Toleration of lowered oxygen tension by cave and stream crayfish. *Ecology*, **29** : 360–367.

71. BURGEN, A. S. V. and KUFFLER, S. W. 1957. Two inhibitory fibres forming synapses with a single nerve cell in the lobster. *Nature, Lond.* **180** : 1490.

72. BURGER, J. W. 1957. The general form of excretion in the lobster, *Homarus*. *Biol. Bull. Woods Hole*, **113** : 207–223.

73. BURGER, J. W. and SMYTHE, C. McC. 1953. The general form of circulation in the lobster, *Homarus*. *J. cell. comp. Physiol.* **42** : 369–383.

74. BURKE, W. 1954. An organ for proprioception and vibration sense in *Carcinus maenas*. *J. exp. Biol.* **31** : 127–138.

75. BURKHARDT, D. 1962. Spectral sensitivity and other response characteristics of single visual cells in the arthropod eye. In : *Biological receptor mechanisms. Symp. Soc. Exp. Biol.* **16** : 86–109.

76. BURTT, E. T. and CATTON, W. T. 1962. The resolving power of the compound eye. In : *Biological receptor mechanisms. Symp. Soc. Exp. Biol.* **16** : 72–85.

77. BUSH, B. M. H. 1962. Peripheral inhibition in the claw of the crab, *Carcinus maenas*. (L). *J. exp. Biol.* **39** : 71–88.

78.   1962. Proprioceptive reflexes in the legs of *Carcinus maenas* (L). *J. exp. Biol.* **39** : 89–105.

78b.   1965. Proprioception by the coxo-basal chordotonal organ, CB, in legs of the crab, *Carcinus maenas*. *J. exp. Biol.* **42** : 285–97.

79. BUSH, B. M. H., WIERSMA, C. A. G. and WATERMAN, T. H. 1964. Efferent mechanoreceptor responses in the optic nerve of the crab *Podophthalmus*. *J. cell. comp. Physiol.* **64** : 327–346.

79b. CALDWELL, P. C. 1964. Calcium and the contraction of *Maia* muscle fibres. *Proc. Roy. Soc. Lond. B.* **160** : 512–516.

80. CALDWELL, P. C. and WALSTER, G. E. 1961. A cannulated crab muscle fibre. *J. Physiol.* **157** : 36P.

81.   1963. Studies on the micro-injection of various substances into crab muscle fibres. *J. Physiol.* **169**, 353–372.

82. CAMIEN, M. N., SARLET, H., DUCHATEAU, G. and FLORKIN, M. 1951. Non-protein amino-acids in muscle and blood of marine and fresh-water crustacea. *J. Biol. Chem.* **193** : 881–885.

83. CAMOUGIS, G. 1964. Visual responses in crayfish. II, Central transmission and interpretation. *J. cell. comp. Physiol.* **63** : 339–352.

84.   1960. Visual responses in crayfish. I, Recording shock responses to light with remote electrodes. *J. cell. comp. Physiol.* **55** : 189–194.

85. CANNON, H. G. 1940. On the anatomy of *Gigantocypris mulleri*. ' *Discovery* ' *Rep.* **19** : 185–244.

86.   1947. On the anatomy of the pedunculate barnacle, *Lithotrya*. *Phil. Trans. Roy. Soc. Lond., B*, **233** ; 89–136

87. CAREY, F. G. and WYATT, G. R. 1960. Uridine diphosphate derivatives in the tissues and haemolymph of insects. *Biochim. biophys. Acta.* **41** : 178–179.

88. CARLISLE, D. B.   1953.   Moulting hormones in *Leander* (Crustacea, Decapoda).   *J. mar. biol. Ass. U.K.* **32** : 289–296.

89.   1953.   Studies on *Lysmata seticaudata*, Risso.   Crustacea, Decapoda.   IV, On the site of the moult-accelerating principle, experimental evidence.   *Pubbl. Staz. zool. Napoli*, **24** : 284–291.

90.   1955.   On the hormonal control of water balance in *Carcinus*. *Pubbl. Staz. zool. Napoli*, **27** : 227–231.

91.   1956.   An indole-alkylamine regulating heart beat in Crustacea. *Biochem. J.* **63** : 32–33.

92.   1957.   On the hormonal inhibition of moulting in decapod Crustacea.   II, The terminal anecdysis in crabs.   *J. mar. biol. Ass. U.K.* **36** : 291–307

92b.   1965.   The effects of Crustacean and Locust ecdysons on moulting and pre-ecdysis in juvenile shore crabs, *Carcinus maenas*. *Gen. comp. Endocrinol.* **5** : 366–372.

92c.   1959.   On the sexual biology of *Pandalus borealis* (Crustacea Decapoda).   III, The inhibition of the female phase.   *J. mar. biol. Assoc.* **38** : 493–506.

93. CARLISLE, D. B. and DOHRN, P. F. R.   1953.   Studies on *Lysmata seticaudata* Risso.   (Crustacea, Decapoda).   II, Experimental evidence for a growth and moult accelerating factor obtainable from eyestalks.   *Pubbl. Staz. zool. Napoli*, **24** : 69.

94. CARLISLE, D. B. and KNOWLES, F. G. W.   1959.   *Endocrine Control in Crustaceans*.   Cambridge University Press.

95. CARLSON, S. P.   1935.   The color changes in *Uca pugilator*.   *Proc. N.Y. Acad. Sci. U.S.* **21** : 549–551.

96. CASE, J.   1964.   Properties of the dactyl chemoreceptors of *Cancer antennarius* Stimpson & *C. productus* Randall.   *Biol. Bull. Woods Hole*, **127** : 428–441.

97. CASE, J., GWILLIAM, G. F. and HANSON, F.   1960.   Dactyl chemoreceptors of Brachyurans.   *Biol. Bull. Woods Hole*, **119** : 308.

98. CASE, J. and GWILLIAM, G. F.   1961.   Amino acid sensitivity of the dactyl chemoreceptors of *Carcinides maenas*.   *Biol. Bull. Woods Hole*, **121** : 449–455.

99. CHAPMAN, G.   1965.   Thesis, Southampton.

100. CHARNIAUX-COTTON, H.   1954.   Découverte chez un Crustacé Amphipode (*Orchestia gammarella*) d'une glande endocrine responsable de la différentiation des charactères sexuels primaires et secondaires mâles.   *Compt. Rend.* **239** : 780–782.

101.   1962.   Androgenic gland of Crustaceans.   *Gen. & Compar. Endocr. Suppl.* **1** : 241–247.

102.   1960.   In : *The Physiology of Crustacea*, ed. Waterman, T. H. Academic Press.

103. CHASSARD, C. 1958. Adaptation chromatique et cycle d'intermue chez *Leander squilla* (Crustacé Decapoda). *Compt. Rend.* **247** : 1039.

104. CLARK, G. P. 1896. On the relation of the otocysts to equilibrium phenomena in *Galasimus pugilator* and *Platyonichus ocellatus*. *J. Physiol.* **19** : 327–343.

105. COHEN, M. J. 1955. The function of receptors in the statocyst of the lobster, *Homarus americanus*. *J. Physiol.* **130** : 9–34.

105b. 1963. Muscle fibres and efferent nerves in a crustacean muscle receptor. *Quart. J. micro. Sci.* **104** : 551–559.

106. COKER, R. E. 1934. Some aspects of the influence of temperature on Copepods. *Science,* **79** : 323.

107. COOKE, I. 1962. Effects of the pericardial organ neurosecretory substance on the crustacean heart. *Gen. & comp. Endocr.* **2** : 29.

108. COOMBS, J. S., ECCLES, J. C. and FATT, P. 1955. The specific ionic conductance and the ionic movements across the motor neuronal membrane that produce the inhibitory post synaptic potential. *J. Physiol.* **130** : 326–373.

109. CONWAY, E. J. 1958. Nature and significance of concentration relations of potassium and sodium ions in skeletal muscle. *Physiol. Rev.* **37** : 84–132.

110. CORI, C. F., CORI, G. T. and GREEN, A. A. 1943. Crystalline muscle phosphorylase. III, Kinetics. *J. Biol. Chem.* **151** : 39–55.

111. CORNER, E. D. S. 1961. On the nutrition and metabolism of zooplankton. *J. mar. biol. Ass. U.K.* **41** : 5–16.

111b. CORNER, E. D. S. and NEWELL, B. S. 1967. On the nutrition and metabolism of zooplankton. The forms of nitrogen excreted by *Calanus*. *J. mar. biol. Assoc.* **47** : 113–120.

112. COSTLOW, J. D. 1963. Moulting and cyclical activity in chromatophototrophins of the central nervous system of the barnacle, *Balanus eburneus*. *Biol. Bull. Woods Hole,* **124** : 254–261.

113. COSTLOW, J. D. and BOOKHOUT, C. G. 1953. Moulting and growth in *Balanus improvisus*. *Biol. Bull. Woods Hole,* **105** : 420–433.

114. 1956. Moulting and shell growth in *Balanus amphitrite niveus*. *Biol. Bull. Woods Hole,* **110** : 107–116.

115. CROGHAN, P. C. 1958. The osmotic and ionic regulation of *Artemia salina*. *J. exp. Biol.* **35** : 219–233.

116. 1958. The mechanism of osmotic regulation in *Artemia salina* (L) : the physiology of the branchiae. *J. exp. Biol.* **35** : 234–242.

117. 1958. The mechanism of osmotic regulation in *Artemia salina* (L) : the physiology of the gut. *J. exp. Biol.* **35** : 243–249.

274　Aspects of the Physiology of Crustacea

118. CROZIER, W. J. and WOLF, E. 1939. The flicker response contour for the crayfish. *J. gen. Physiol.* **23** : 1–10.

119. DAINTY, J. 1962. Ion transport and electrical potentials in plant cells. *A. Rev. Pl. Physiol.* **13** : 379–402.

120. DAMBOVICEAUNU, A. 1932. *Arch. Roum. Path. Exp. Microbiol.* **5** : 239–309. (Quoted in 510).

121. DAUMER, K., JANDER, R. and WATERMAN, T. H. 1963. Orientation of the ghost crab, *Ocypode*, in polarized light. *Z. vergl. Physiol.* **47** : 56–76.

122. DAVIS, E. and BURNETT, A. L. 1964. A study of growth and cell differentiation in the hepatopancreas of the crayfish. *Develop. Biol.* **10** : 122–153.

123. DAVIS, F. and NACHMANSOHN, D. 1964. Acetylcholine formation in lobster sensory axons. *Biochim. biophys. Acta*, **88** : 384–389.

124. DEAN, J. M. and VERNBERG, F. J. 1965. Variations in the blood glucose level of crustacea. *Comp. Biochem. Physiol.* **14** : 29.

125. DE BRUIN, G. H. P. and CRISP, D. J. 1957. The influence of pigment migration on vision of higher crustacea. *J. exp. Biol.* **34** : 447–463.

126. DEHNEL, P. A. 1960. Effect of temperature and salinity on the oxygen consumption of two intertidal crabs. *Biol. Bull. Woods Hole*, **118** : 215–249.

127. ——— 1960. Aspects of osmoregulation in two species of intertidal crabs. *Biol. Bull. Woods Hole*, **122** : 208–227.

128. DEHNEL, P. A. and CAREFOOT, T. H. 1965. Ion regulation in two species of intertidal crabs. *Comp. Biochem. Physiol.* **15** : 377–397.

129. DELAUNAY, H. 1934. Le métabolisme de l'ammoniaque d'après les recherches relatives aux invertébrés. *Annls Physiol. Physiochim. biol.* **10** : 695–729.

130. DEL CASTILLO, J. and KATZ, B. 1953. Statistical nature of ' facilitation ' at a single nerve muscle junction. *Nature, Lond.* **171** : 1016–1017.

131. DEMEUSY, N. 1958. Recherches sur la mue de puberté du Decapoda Brachyoure, *Carcinus maenas*. *Archs Zool. exp. gen.* **95** : 253.

132. ——— 1960. *Cah. Biol. mar.* **1** : 257–277.

133. ——— 1962. *Cah. Biol. mar.* **3** : 37–56.

134. DEMEUSY, N. and VEILLET, A. 1958. Influence de l'ablation des pédoncles oculaires sur la glande androgène de *Carcinus maenas* (L). *C. R. Acad. Sci. Paris*, **246** : 1104–1107.

135. DEMEUSY, N., CORNUBERT, G. and VEILLET, A. 1952. Effets de l'ablation des pédoncles oculaires sur le développement des charactères sexuels externes des Decapodes Brachyoures *Carcinus maenas* Pennant et *Pachygrapsus marmoratus* Fabricus. *C. R. Acad. Sci. Paris*, **234** : 1405–1407.

136. DENNELL, R. 1937. On the feeding mechanism of *Apseudes talpa* and the evolution of the pericaridan feeding mechanisms. *Trans. R. Soc. Edinb.* **59** : 57–78.

137. —— 1947. The occurrence and significance of phenolic hardening in newly formed cuticle of Crustacea Decapoda. *Proc. R. Soc. Lond.*, *B*, **134** : 485–503.

138. DETHIER, V. G. 1962. Chemoreceptive mechanisms in insects. *Symp. Soc. exp. Biol.* **16** : 180–196.

139. DEVILLEZ, E. J. 1965. Isolation of the proteolytic digestive enzymes from the gastric juice of the crayfish, *Orconectes virilis* (Hagen). *Comp. Biochem. Physiol.* **14** : 577–586.

140. DIGBY, P. B. S. 1961. Mechanism of sensitivity to hydrostatic pressure in the prawn, *Palaemonetes varians*. *Nature*, **191** : 366–367.

141. DIJKGRAAF, S. Quoted by Cohen in : *The Physiology of Crustacea* ed. Waterman, T. H. Academic Press.

142. DOBKIN, S. and MANNING, R. B. 1964. Osmoregulation in two species of *Palaemonetes* (Crustacea : Decapoda) from Florida. *Bull. mar. Sci. Gulf Caribb.* **14** : 149–157.

143. DORAI RAJ, B. S. 1964. Functional separation of crab muscle fibres innervated by a single motor axon. *J. cell. comp. Physiol.* **64** : 41–54.

144. DRACH, P. 1939. Mue et cycle d'intermue chez les Crustacés décapodes. *Annls Inst. océanogr.*, Monaco, **19**, 103–391.

145. DRESEL, E. I. B. and MOYLE, V. 1950. Nitrogenous excretion of amphipods and isopods. *J. exp. Biol.* **27** : 210–225.

146. DRESEL, E. I. B. 1948. Passage of haemoglobin from blood into eggs of *Daphnia*. *Nature*, **162** : 736.

147. DRILHON, H. 1933. La glucose et la mue des crustacés. *C. R. Acad. Sci. Paris*, **196** : 506–508.

148. DUBUISSON. Quoted in 72.

149. DUCHATEAU, G. and FLORKIN, M. 1955. Concentration du milieu extérieur et état stationnaire du pool des acides aminés non-protéiques des muscles. *Archs int. Physiol.* **63** : 249–251.

150. DUCHATEAU, G., FLORKIN, M. and JEUNIAUX, C. 1959. Composante amino acide des tissus chez les Crustacés. I, Composante amino acide des muscles de *Carcinus maenas* L. lors du passage de l'eau saumâtre et au cours de la mue. *Archs int. Physiol.* **67** : 489–500.

151. DUCHATEAU-BOSSON, G., JEUNIAUX, C. and FLORKIN, M. 1961. Rôle de la variation de la composante amino-acide intracellulaire dans l'euryhalinité d'*Arenicola marina* L. *Archs int. Physiol.* **69** : 30–35.

152. —— 1961. Change in intracellular concentration of free amino acids as a factor of euryhalinity in the crayfish, *Astacus astacus*. *Comp. Biochem. Physiol.* **3** : 245–250.

153. DUCHATEAU-BOSSON, G. and FLORKIN, M. 1962. Adaptation à l'eau de mer de crabes chinois (*Eriocheir sinensis*) présentant dans l'eau douce une valeur élevée de la composante amino acide des muscles. *Archs int. Physiol.* **70** : 345–355.

154. DUDEL, J. and KUFFLER, S. W. 1961. The quantal nature of transmission and spontaneous miniature potentials at the crayfish neuromuscular junction. *J. Physiol.* **155** : 514–529.

155. —— 1961. Mechanisms of facilitation at the crayfish neuromuscular junction. *J. Physiol.* **155** : 530–542.

156. —— 1961. Presynaptic inhibition at the crayfish neuromuscular junction. *J. Physiol.* **155** : 543–562.

157. DUDEL, J. GRYDER, R., KAYI, A., KUFFLER, S. W. and POTTER, D. D. 1963. Gamma-aminobutyric acid and other blocking compounds in Crustacea. I, Central nervous system. *J. Neurophysiol.* **26** : 721–728.

158. DURAND, J. B. 1956. Neurosecretory cell types and their activity in the crayfish. *Biol. Bull. Woods Hole*, **111** : 62–76.

159. DURAND, J. B. 1960. Limb regeneration and endocrine activity in the crayfish. *Biol. Bull. Woods Hole*, **118** : 250–261.

160. DUVAL, M. 1925. Recherches physico-chimiques et physiologiques sur le milieu intérieur des animaux aquatiques. *Ann. Inst. Océanogr. Monaco, N.S.* **2** : 233–407.

161. EASTON, D. M. 1950. Synthesis of acetylcholine in crustacean nerve and nerve extract. *J. Biol. Chem.* **185** : 813–816.

162. ECHALIER, G. 1955. Rôle de l'organe Y dans le déterminisme de la mue de *Carcinides* (*Carcinus*) *maenas* (L) Crustacés décapodes. Expériences d'implantation. *Compt. Rend. Soc. Biol. Paris*, **240** : 1581–1583.

162b. ECKERT, R. O. 1961. Reflex relationships of the abdominal stretch receptors of the crayfish. II, Stretch receptor involvement during the swimming reflex. *J. cell. comp. Physiol.* **57** : 163–174.

163. EDLÉN quoted in 221.

164. EDNEY, E. B. 1951. The body temperature of woodlice. *J. exp. Biol.* **30** : 331–349.

165. —— 1951. The evaporation of water from woodlice and the millipede, *Glomeris*. *J. exp. Biol.* **28** : 91–115.

166. EDNEY, E. B. 1954. Woodlice and the land habitat. *Biol. Rev.* **29** : 185–219.

167. EDWARDS, C. and HAGIWARA, S. 1959. Potassium ions and the inhibitory process in the crayfish stretch receptor. *J. gen. Physiol.* **43** : 315–321.

168. EDWARDS, C. and KUFFLER, S. W. 1959. The blocking effect of gamma amino butyric acid (GABA) and the action of related compounds on single nerve cells. *J. Neurochem.* **4** : 19–30.

169. EDWARDS, C. and OTTOSON, D. 1958. The site of impulse initiation in a nerve cell of a crustacean stretch receptor. *J. Physiol.* **143** : 138–148.

170. EDWARDS, G. A. 1950. The influence of eyestalk removal on the metabolism of the fiddler crab. *Physiologia comp. Oecol.* **2** : 34–50.

171. ELLIOTT, K. A. C. and FLOREY, E. 1956. Factor 1—Inhibitory factor from brain. Assay, condition in brain. Stimulating and antagonistic substances. *J. Neurochem.* **1** : 181–192.

172. ENRIGHT, J. T. 1962. Response of an amphipod to pressure changes. *Comp. Biochem. Physiol.* **7** : 131–146.

173. EWER, D. W. and HATTINGH, I. 1952. Absorption of silver ions on the gills of a fresh-water crab. *Nature*, **169** : 460.

174. EYZAGUIRRE, C. 1961. Excitatory and inhibitory processes in Crustacea sensory nerve cells. In : *Nervous Inhibition* ed. Florey, E. Pergamon.

175. EYZAGUIRRE, C. and KUFFLER, S. W. 1955. Processes of excitation in the dendrites and in the soma of single isolated sensory nerve cells of the lobster and crayfish. *J. gen. Physiol.* **39** : 87–119.

176. ——— 1955. Further study of soma, dendrite and axon excitation in single neurones. *J. gen. Physiol.* **39** : 120–153.

177. FARRELL, G. and TAYLOR, A. N. 1962. Neuroendocrine aspects of blood volume regulation. *A. Rev. Physiol.* **24** : 471–490.

178. FATT, P. and KATZ, B. 1953. The electrical properties of crustacean muscle fibres. *J. Physiol.* **120** : 171–204.

179. ——— 1952. Spontaneous sub-threshold activity in motor nerve endings. *J. Physiol.* **117** : 109–128.

180. ——— 1953. The effect of inhibitory nerve impulses on a crustacean muscle fibre. *J. Physiol.* **121** : 374–389.

180b. FIELDS, H. L. 1966. Proprioceptive control of posture in the crayfish abdomen. *J. exp. Biol.* **44** : 455–469.

181. FINGERMAN, M. 1955. The rate of oxygen consumption of the dwarf crayfish, *Cambarellus shufeldtii. Anat. Rec.* **122** : 464.

182. FINGERMAN, M.   1957.   Physiology of the red and white chromato-phores of the dwarf crayfish, *Cambarellus shufeldtii*. *Physiol. Zool.* **30** : 142–154.

183.   1965. Chromatophores. *Physiol. Rev.* **45** : 296–339.

184. FINGERMAN, M. and AOTO, T.   1962.   Regulation of pigmentary phenomena.   Hormonal regulation of pigmentary effectors in Crustaceans.   *Gen. & comp. Endocr. Suppl.* **1** : 81–93.

185. FINGERMAN, M. and MOBBERLY, W. C. JR.   1960.   Investigation of the hormones controlling the distal retinal pigment of the prawn, *Palaemonetes*. *Biol. Bull. Woods Hole*, **118** : 393–406.

186. FINGERMAN, M., LOWE, M. E. and SUNDARARAJ, B. I.   1959. Dark adapting and light adapting hormones controlling the distal retinal pigment of the prawn, *Palaemonetes vulgaris*. *Biol. Bull. Woods Hole*, **116** : 30–36.

187. FLEMISTER, L. J.   1958.   Salt and water anatomy, constancy and regulation in related crabs from marine and terrestrial habitats. *Biol. Bull. Woods Hole*, **115** : 180–200.

188. FLEMISTER, L. J. and FLEMISTER, S. C.   1951.   Chloride ion regulation and oxygen consumption in the crab *Ocypode albicans* (Bosq.).   *Biol. Bull. Woods Hole*, **101** : 259–273.

189. FLEMISTER, S. C.   1959.   Histophysiology of gill and kidney of the crab, *Ocypode albicans*. *Biol. Bull. Woods Hole*, **116** : 37–48.

190. FLOREY, E.   1954.   *Archs int. Physiol.* **62** : 33.

191.   1957.   Chemical transmission and adaptation. *J. gen. Physiol.* **40** : 533–545.

192.   1960.   Physiological evidence for naturally occurring inhibitory substances.   In : *Inhibition in the Nervous System and Gamma-aminobutyric acid.*   72–84.   Pergamon Press.

193.   1964.   Amino-acids as transmitter substances. In : *Major problems of Neurendocrinology.* ed. Boujusy, E. and Jasmin, G., Karger. N.Y.

194. FLOREY, E. and CHAPMAN, D. D.   1961.   The non-identity of the transmitter substance of crustacean inhibitory neurons and gamma-aminobutyric acid.   *Comp. Biochem. Physiol.* **3** : 92–98.

195. FLOREY, E. and FLOREY, E.   1955.   Microanatomy of the abdominal stretch receptors of the crayfish (*Astacus fluviatilis* L.) *J. gen. Physiol.* **39** : 69–85.

196. FLORKIN, M.   1936.   Sur le taux de la glycémie plasmatique vrai chez les crustacés décapodes. *Bull. Acad. r. Belge Cl. Sci.* **22** : 1359–1367.

197.   1962.   La régulation isosmotique intracellulaire chez les invertébrés marins euryhalins.   *Bull. Acad. r. Belge Cl. Sci.* **48** : 687–694.

198. FLORKIN, M. and DUCHATEAU, G. 1939. *C. R. Soc. Biol.* **132** : 484–486.

199. FORSTER, R. and ZIA WOHLRATH, P. 1941. The absence of active secretion as a factor in the elimination of inulin and other substances by the green glands of the lobster *Homarus americanus*. *Anat. Rec.* **81** : *Suppl.* 128.

200. FOX, H. M. 1939. The activity and metabolism of poikilothermic animals in different latitudes. V. *Proc. zool. Soc. Lond.* **109** *A* : 141–156.

201. ——— 1945. The oxygen affinities of certain invertebrate haemoglobins. *J. exp. Biol.* **21** : 161–165.

202. ——— 1948. The haemoglobin of *Daphnia*. *Proc. R. Soc., B,* **135** : 195.

203. ——— 1949. On *Apus* : its rediscovery in Britain, nomenclature and habits. *Proc. zool. Soc. Lond.* **119** : 693–702.

204. ——— 1952. Anal and oral intake of water by crustacea. *J. exp. Biol.* **29** : 583–599.

205. FOX, H. M. and JOHNSON, M. L. 1934. The control of respiratory movements in crustacea by oxygen and carbon dioxide. *J. exp. Biol.* **11** : 1–10.

206. FOX, H. M. and PHEAR, E. A. 1953. Factors influencing haemoglobin synthesis in *Daphnia*. *Proc. R. Soc., B.* **141** : 179–189.

207. FOX, H. M. and SIMMONDS, B. G. 1933. Metabolic rates of aquatic arthropods from different salinities. *J. exp. Biol.* **10** : 67–78.

208. FOX. H. M. and WINGFIELD, C. A. 1937. The activity and metabolism of poikilothermic animals in different latitudes. II. *Proc. zool. Soc. Lond. A,* **107** : 275–282.

209. FOX, H. M., GILCHRIST, B. M. and PHEAR, E. A. 1951. Functions of haemoglobin in *Daphnia*. *Proc. R. Soc., B,* **138** : 514–528.

210. FOX, H. M., HARDCASTLE, S. M. and DRESEL, E. I. B. 1949. Fluctuations in the haemoglobin content of *Daphnia*. *Proc. R. Soc., B,* **136** : 388–399.

211. FOX, H. M., SIMMONDS, B. G. and WASHBOURN, R. 1935. Metabolic rates of Ephemerid nymphs from swiftly flowing and from still waters. *J. exp. Biol.* **12** : 179–184.

212. FRITSCHE, H. 1916. Studien über die schwankungen des osmotischen Druckes der Körperflussigkeit bei *Daphnia magna*. *Int. Rev. ges. Hydrobiol.* **8** : 125–203.

213. FURSHPAN, E. J. and POTTER, D. D. 1959. Transmission at the giant motor synapses of the crayfish. *J. Physiol.* **145** : 289–325.

214. ——— 1959. Slow post synaptic potentials recorded from the giant motor fibres of the crayfish. *J. Physiol.* **145** : 326–335.

215. GABE, M. 1953. Sur l'existence chez quelques Crustacés Malacostraces d'une organe comparable à la gland de la mue des insectes. *C. R. Soc. Biol. Paris*, **237** : 1111–1113.

216.   1956. Histologie comparée de la glande de mue (organe Y) des Crustacés Malacostraces. *Annls Sci. nat. Zool. Biol. Animale*, **18** : 145–152.

217. GARSTANG, W. 1896. The habits and respiratory mechanisms of *Corystes cassivelaunus*. *J. mar. biol. Ass. U.K.* **4** : 223–232.

218. GAULD, D. T. 1953. Diurnal variations in the grazing of planktonic copepods. *J. mar. biol. Ass. U.K.* **31** : 461–474.

218b. GEORGE, R. Y. 1966. Glycogen content in the wood-boring isopod *Limnoria lignorum*. *Science*, **153** : 1262–1264.

219. GILCHRIST, B. M. 1956. The oxygen consumption of *Artemia salina* (L) in different salinities. *Hydrobiologica*, **8** : 54–65.

220. GIRARDIER, L., REUBEN, J. P., BRANDT, P. W. and GRUNDFEST, H. 1963. Evidence for anion-permeable membrane in crayfish muscle fibres and its possible role in excitation-contraction coupling. *J. gen. Physiol.* **47** : 189–214.

221. GNANAMUTHU, C. P. 1948. Notes on the anatomy and physiology of *Caligus savala* n. sp., a parasitic copepod from Madras plankton. *Proc. zool. Soc. Lond.* **118** : 591–606.

222. GRAY, I. E. 1957. A comparative study of gill area of crabs. *Biol. Bull. Woods Hole*, **112** : 34–42.

223. GREEN, J. 1956. Growth, size and reproduction in *Daphnia*. (Crustacea : Cladocera). *Proc. zool. Soc. Lond.* **126** : 173–204.

224.   1961. *A Biology of Crustacea*. H. F. and G. Witterby Ltd., London.

224b. GREEN, J. P. 1964. Morphological colour change in the fiddler crab *Uca pugnax* (S. I. Smith). *Biol. Bull. Woods Hole*, **127** : 239–255.

225. GREEN, J. W., HARSCH, M., BARR, L. and PROSSER, C. L. 1959. The regulation of water and salt by the fiddler crabs, *Uca pugnax* and *Uca pugilator*. *Biol. Bull. Woods Hole*, **116** : 76–87.

226. GROSS, W. J. 1955. Aspects of osmoregulation of crabs showing the terrestrial habit. *Am. Nat.* **89** : 205–222.

227.   1957. An analysis of response to osmotic stress in selected decapod crustacea. *Biol. Bull. Woods Hole*, **112** : 43–62.

228.   1957. A behavioural mechanism for osmotic regulation in a semi-terrestrial crab. *Biol. Bull. Woods Hole*, **113** : 268–274.

229.   1958. Potassium and sodium regulation in an inter-tidal crab. *Biol. Bull. Woods Hole*, **114** : 334–347.

230. GROSS, W. J.   1959.   The effect of osmotic stress on the ionic exchange of a shore crab.   *Biol. Bull. Woods Hole*, **116** : 248–257.

231.   1963.   Cation and water balance in crabs showing the terrestrial habit.   *Physiol. Zool.* **36** : 312–324.

232.   1964.   Trends in water and salt regulation among aquatic and amphibious crabs.   *Biol. Bull. Woods Hole*, **127** : 447–466.

233.   1964.   Water balance in anomuran land crabs on a dry atoll.   *Biol. Bull. Woods Hole*, **126** : 54–68.

234. GROSS, W. J. and HOLLAND, P.   1960.   Water and ionic regulation in a terrestrial hermit crab.   *Physiol. Zool.* **33** : 21–28.

235. GROSS, W. J. and MARSHALL, L. A.   1960.   The influence of salinity on the water fluxes of a crab.   *Biol. Bull. Woods Hole*, **119** : 440–453.

236. HAGIWARA, S. and NAKA, K.   1964.   The initiation of spike potentials in barnacle muscle fibres under low intra-cellular Ca^{++}.   *J. gen. Physiol.* **48** : 141–162.

237. HAGIWARA, S., CHICHIBU, S. and NAKA, K.   1964.   The effects of various ions on resting and spike potentials of barnacle muscle fibres.   *J. gen. Physiol.* **48** : 163–179.

238. HAGIWARA, S., KUSANO, K. and SAITO, S.   1960.   Membrane changes in crayfish stretch receptor neurone during synaptic inhibition and under action of gamma-aminobutyric acid.   *J. Neurophysiol.* **23** : 505–514.

239. HARDY, W. B.   1894.   On some histological features and physiological properties of the post-oesophageal nerve cord of the crustacean.   *Phil. Trans. r. Soc. B.* **185** : 83–117.

240. HARTENSTEIN, R.   1964.   Feeding, digestion, glycogen and the environmental conditions of the digestive system in *Oniscus asellus*.   *Insect Physiol.* **10** : 611–621.

241. HARTLINE, H. K.   1928.   A quantitative and descriptive study of the electric response to illumination of the arthropod eye.   *Am. J. Physiol.* **83** : 466–483.

242. HARTOG, M. M.   1879.   On *Cyclops*.   *Rep. Brit. Ass.* 374.

243. HARVEY, H. W.   1937.   Notes on selective feeding by *Calanus*.   *J. mar. biol. Ass. U.K.* **22** : 97–100.

244. HARVEY, H. W.   1942.   Production of life in the sea.   *Biol. Rev.* **17** : 221–246.

244b. HASSELBACH, W. and WEBER, H. H.   1965.   Die intrazelluläre Regulation der Muskelaklivität.   *Naturwissenschaften* **52** : 121–128.

245. HEILBRUNN, L. V.   1940.   Action of calcium on muscle protoplasm.   *Physiol. Zool.* **13** : 88–94.

246. HEILBRUNN, L. V. and WIERCINSKI, F. J.   1947.   The action of various cations on muscle protoplasm.   *J. cell. comp. Physiol.* **29** : 15–32.

247. HELFF, O. M.   1928.   The respiratory regulation of the crayfish, *Cambarus immunis* (Hagen).   *Physiol. Zool.* **1** : 76–96.

248. HEMMINGSEN, A. M.   1924.   Blood sugar regulation in the crayfish.   *Skand. Arch. Physiol.* **46** : 51–55.

249. HERMANN, H. T.   1964.   Stochastic properties in the negative phototropic behaviour of the crayfish.   *J. exp. Zool.* **155** : 381–401.

250. HERRNKIND, W. F. and CUMMINGS, W. C.   1964.   Single file migration in the spiny lobster, *Panulirus argus* (Latreille).   *Bull. mar. Sci. Gulf Caribb.* **14** : 123–125.

251. HICKMAN, C. P.   1959.   The osmoregulatory role of the thyroid in the starry flounder *Platichthys stellatus*.   *Canad. J. Zool.* **37** : 997–1060.

252. HIESTAND, W. A.   1931.   The influence of varying tensions of oxygen upon the respiratory metabolism of certain aquatic insects and crayfish.   *Physiol. Zool.* **4** : 246–270.

253. HODGKIN, A. L.   1951.   The ionic basis of electrical activity in nerve and muscle.   *Biol. Rev.* **26** : 339–409.

254. HODGKIN, A. L. and HOROWICZ, P.   1957.   The differential action of hypertonic solutions on the twitch and action potential of a muscle fibre.   *J. Physiol.* **136** : 17 pp.

255. HODGKIN, A. L. and KATZ, B.   1949.   The effect of sodium ions on the electrical activity of the giant axon of the squid.   *J. Physiol.* **108** : 37–77.

256. HODGSON, E. S.   1958.   Electrophysiological studies of Arthropod chemoreception.   I, Chemoreception of terrestrial and freshwater arthropods.   *Biol. Bull. Woods Hole*, **115** : 114–125.

257. HAEFNER, P. A. and SHUSTER, C. N.   1964.   Length increments during terminal moult of the female blue crab, *Callinectes sapidus* in different salinity environments.   *Chesapeake Sci.* **5** : 114–118.

258. HOGBEN, L. T.   1926.   Some observations on the dissociation of haemocyanin by the colorimetric method.   *J. exp. Biol.* **3** : 225–238.

259. HOLMES, W., PUMPHREY, R. J. and YOUNG, J. Z.   1941.   The structure and conduction velocity of the medullated nerve fibres of prawns.   *J. exp. Biol.* **18** : 50–54.

260. HOYLE, G.   1955.   *Proc. Roy. Soc. Lond. B*, **143** : 281–292.

261. HOYLE, G. and WIERSMA, C. A. G.   1958.   Excitation at neuromuscular junctions in crustacea.   *J. Physiol.* **143** : 403–425.

262. HOYLE, G. and WIERSMA, C. A. G.   1958.   Inhibition at neuro-muscular junctions in crustacea.   *J. Physiol.* **143** : 426–440.

263.   1958.   Coupling of membrane potential to contraction in crustacean muscles.   *J. Physiol.* **143** : 441–453.

264. HRABE, S.   1949.   O reduci $AgNO_3$ a $KMnO_4$ na urcitych mistech tela koregsu *Asellus aquaticus* a *Synurella ambulans*.   *Zpr. Anthrop. Spol.* **11** : 1–4.

265. HU, A. S. L.   1958.   Glucose metabolism in the crab *Hemigrapsus nudus*.   *Archs Biochem. Biophys.* **75** : 387–395.

266. HUBSCHMAN, J. H.   1963.   Development and function of neuro-secretory sites in the eyestalks of *Palaemonetes*.   *Biol. Bull. Woods Hole*, **125** : 96–113.

266b. HUGGINS, A. K.   1966.   Contribution to *The nature and function-ing of enzymes in organisms adapted to the aquatic habitat*.   Fourth meeting on the biochemistry of aquatic organisms, Southampton, 1965.

266c.   1966.   Intermediary metabolism in *Carcinus maenas*.   *Comp. Biochem. Physiol.* **18** : 283–290.

267. HUGHES, G. M. and WIERSMA, C. A. G.   1960.   The co-ordination of swimmeret movements in the crayfish, *Procambarus clarkii* (Girard).   *J. exp. Biol.* **37** : 657–670.

268.   1960.   Neuronal pathways and connections in the abdominal cord of the crayfish.   *J. exp. Biol.* **37** : 291–307.

269. HURLEY, D.   1959.   Notes on the ecology and environmental adaptations of the terrestrial Amphipoda.   *Pacif. Sci.* **13** : 107–129.

270. HUSSON, R. and HENRY, J. P.   1963.   Étude chez trois espèces du genre *Asellus* des concrétions des organes de Zenker.   *Bull. Soc. zool. Fr.* **88** : 274–277.

271. HUTTER, O. F. and TRAUTWEIN, W.   1955.   Effect of vagal stimulation on the sinus venosus of the frog's heart.   *Nature*, **176** : 512–3.

272. HUXLEY, A. F. and TAYLOR, R. E.   1958.   Local activation of striated muscle fibres.   *J. Physiol.* **144** : 426–441.

273. HUXLEY, T. H.   1906.   *The Crayfish*.   London, Kegan Paul; Trench.

274. HYNES, H. B. N.   1954.   The ecology of *Gammarus duebeni*, Lilljeborg, and its occurrence in fresh water in Western Britain.   *J. Anim. Ecol.* **23** : 38–84.

275. JENKIN, P. A.   1962.   *Animal Hormones ; a Comparative Survey*.   Pergamon Press.

276. Jeuniaux, C. 1956. Chitinase et bactéries chitinolytiques dans le tube digestif d'une claporte, (*Porcellio scaber* Latr.) (Isopode, Oniscide). *Archs int. Physiol. Biochim.* **64** : 583–586.

277. Jeuniaux, C. and Florkin, M. 1961. Modification de l'excrétion azotée du crabe chinois au cours de l'adaption osmotique. *Archs int. Physiol.* **69** : 385–386.

278. Jeuniaux, C., Bricteux-Gregoire, S. and Florkin, M. 1962. Régulation osmotique intracellulaire chez *Astacus rubens* glycolle et de la taurine. *Cah. Biol. mar.* **3** : 107–113.

279. Jirovic, O. and Wenig, K. 1934. Quoted by Wolvekamp, H. P. and Waterman, T. H. in : *The Physiology of Crustacea.* Academic Press.

280. Jones, D. 1964. Personal communication.

281. Jones, L. L. 1941. Osmotic regulation in several crabs of the Pacific coast of North America. *J. cell. comp. Physiol.* **18** : 79–91.

282. Johnson, G. E. 1924. Giant nerve fibres in crustaceans with special reference to *Cambarus* and *Palaemonetes.* *J. cell. comp. Neurol.* **36** : 323–373.

283. Johnson, M. L. 1936. The control of respiratory movements in Crustacea by oxygen and carbon dioxide. II. *J. exp. Biol.* **13** : 467–475.

284. Kamemoto, F. I., Keister, S. M. and Spalding, A. E. 1962. Cholinesterase activity and sodium movement in the crayfish kidney. *Comp. Biochem. Physiol.* **7** : 81–87.

285. Kao, C. Y. 1960. Postsynaptic electrogenesis in septate giant axons. II, Comparison of medial and lateral giant axons in crayfish. *J. Neurophysiol.* **23** : 618–635.

286. Karlson, P. 1956. *Annls Sci. nat. Zool. Biol. Animale.* **18** : 125–138.

286b. Karlson, P. and Skinner, D. M. 1960. Attempted extraction of crustacean moulting hormone from isolated y-organ. *Nature,* **185** : 543–544.

287. Katz, B. 1936. Neuromuscular transmission in crabs. *J. Physiol.* **87** : 199–220.

288.   1958. *Bull. John Hopkins Hosp.* **102** : 275–295.

288b. Kennedy, D. 1963. Physiology of photoreceptor neurones in the abdominal nerve cord of the crayfish. *J. gen. Physiol.* **46** : 551–572.

288c.   1966. The comparative physiology of invertebrate neurones *Adv. Comp. Physiol. Biochem.* **2** : 117–184.

289. Kennedy, D. and Preston, J. B. 1960. Activity patterns of interneurones in the caudal ganglion of the cray-fish. *J. gen. Physiol.* **43** : 655–670.

289b. KENNEDY, D. and TAKEDA, K.   1965.   Reflex control of abdominal flexor muscles in the crayfish.   I, The twitch system.   *J. exp. Biol.* **43** : 211–227.

289c.   1965.   Reflex control of abdominal flexor muscles in the crayfish.   II. The tonic system.   *J. exp. Biol.* **43** : 229–246.

289d. KENNEDY, D., EVOY, W. H. and HANAWALT, J. T.   1966.   Release of co-ordinated behaviour in crayfish by single central nervous neurones.   *Science,* **154** : 917–919.

290. KERKUT, G. A. and PRICE, M. A.   1963.   Chromatographic separation of cardiaccelerators (6 HT and a mucopeptide) from *Carcinus* heart.   *Comp. Biochem. Physiol.* **11** : 45–52.

291. KEYNES, R. D. and LEWIS, P. R.   1951.   The resting exchange of radioactive potassium in crab nerve.   *J. Physiol.* **113** : 73–98.

292. KINCAID, F. D. and SCHEER, B. T.   1952.   Hormonal control of metabolism in crustaceans.   IV, Relation of tissue composition of *Hemigrapsus nudus* to intermoult cycle and sinus gland.   *Physiol. Zool.* **25** : 372–380.

293. KING, D. S.   1964.   Fine structure of the androgenic gland of the crab, *Pachygrapsus crassipes.*   *Gen. & comp. Endocr.* **4** : 533–544.

294. KINNE, O.   1959.   Ecological data on the amphipod *Gammarus duebeni.*   A monograph.   *Veröff. Inst. Meeresforsch. Bremerh.* **6** : 177–202.

295.   1960.   *Gammarus salinus-* Einige daten über den Umwelteinfluss auf Wachstum, Hautugsolge, Herzfrequenz und Eientwichlungsdaur.   *Crustaceana,* **1** : 208–217.

296. KINNE, O. and ROTTHAUWE, H. W.   1952.   Biologische Beobachtungen und Untersuchungen über die Blutkonzentration an *Heteropanope tridentatus* Maitland (Decapoda). *Kieler Meeresforsch.* **8** : 212–217.

296b. KIRSCHNER, L. B. and WAGNER, S.   1965.   The site and permeability of the filtration locus in the crayfish antennal gland.   *J. exp. Biol.* **43** : 385–395.

297. KLEINHOLTZ, L. H.   1936.   Crustacean eye-stalk hormone and retinal pigment migration.   *Biol Bull. Woods Hole,* **70** : 159–184.

298.   1937.   Colour changes and diurnal rhythm in *Ligia baudiniana.* *Biol. Bull. Woods Hole,* **72** : 24–36.

299.   1960.   Pigment effectors.   In : *The Physiology of the Crustacea.* ed. Waterman, T. H.   Academic Press.

300. KLEINHOLTZ, L. H. and LITTLE, B. C.   1949.   Studies in the regulation of blood sugar concentration in crustaceans.   I, Normal values and experimental hyperglycaemia in *Libinia emarginata.* *Biol. Bull. Woods Hole,* **96** : 218–227.

301. KLEINHOLTZ, L. H., HAVEL, V. J. and REICHART, R. 1950. Studies in the regulation of blood sugar concentration in crustaceans. II, Experimental hyperglycaemia and the regulatory mechanisms. *Biol. Bull. Woods Hole*, **99** : 454–468.

302. KLOTZ, I. M. and KLOTZ, T. A. 1956. The chemical nature of bound oxygen in haemerythrin and haemocyanin. *Biol. Bull. Woods Hole*, **111** : 306.

303. KNOWLES, F. G. W. and CARLISLE, D. B. 1956. Endocrine control in the crustacea. *Biol. Rev.* **31** : 396–473.

304. KOOIMAN, P. 1964. The occurrence of carbohydrases in digestive juice and in hepatopancreas of *Astacus fluviatilis*, Fabr. and of *Homarus vulgaris* M-E. *J. cell. comp. Physiol.* **63** : 197–201.

305. KRAVITZ, E. A., KUFFLER, S. W. and POTTER, D. D. 1963. Gamma-aminobutyric acid and other blocking compounds in the crustacea. III, Their relative concentrations in separated motor and inhibitory axons. *J. Neurophysiol.* **26** : 739–751.

306. KREIDL, A. 1893. Quoted by Clark (1896). *J. Physiol.* **19** : 327–343.

307. KRISHNAN, G. 1951. Phenolic tanning and pigmentation of the cuticle in *Carcinus maenas*. *Q. J. Microsc. Sci.* **92** : 333–344.

308. KRNJEVIC, K. and VAN GELDER, N. M. 1961. Tension changes in crayfish stretch receptors. *J. Physiol.* **159** : 310–325.

309. KROGH, A. 1939. *Osmotic and ionic regulation in aquatic animals* (pp. 1–233). Cambridge Univ. Press.

310. KUFFLER, S. W. 1954. Mechanisms of activation and motor control of stretch receptors of lobster and crayfish. *J. Neurophysiol.* **17** : 558.

311. KUFFLER, S. W. and EDWARDS, C. 1958. Mechanism of gamma-aminobutyric acid (GABA) action and its relation to synaptic inhibition. *J. Neurophysiol.* **21** : 589–610.

312. KUFFLER, S. W. and EYZAGUIRRE, C. 1955. Synaptic inhibition in an isolated nerve cell. *J. gen. Physiol.* **39** : 155–184.

313. KUFFLER, S. W. and KATZ, B. 1946. Inhibition at the nerve muscle junction in Crustacea. *J. Neurophysiol.* **9** : 337–346.

314. KUIPER, J. W. 1962. The optics of the compound eye. In : *Biological Receptor Mechanisms*. *Symp. Soc. exp. Biol.* **16** : 58–71.

315. KUMMEL, G. 1964. Das Coelömosäcken der Antennendrüse von *Cambarus affinis* Say (Decapoda, Crustacea). *Zool. Beitr.* **10** : 227–252.

316. KURUP, N. G. 1964. The incretory organs of the eyestalk and brain of the porcelain crab. *Petrolisthes cintipes* (Reptantia, Anomura). *Gen. & Comp. Endocr.* **4** : 99–112.

316b. KUSANO, K. 1966. Electrical activity and structural correlates of giant nerve fibres in Kuruma shrimp (*Penaeus japonicus*). *J. cell. comp. Physiol.* **68** : 361–383.

317. LAGERSPETZ, K. and MATTILA, M. 1961. Salinity reactions of some fresh and brackish water crustaceans. *Biol. Bull. Woods Hole,* **120** : 44–53.

318. LANCASTER, E. R. 1871. Über das Vorkommen von Haemoglobin in den Muskeln der Mollusken und die Verbreitung desselben in den lebendigen Organismen. *Pflügers. Arch. ges. Physiol.* **4** : 315.

319. LANCE, J. 1963. The salinity tolerance of some estuarine planktonic crustaceans. *Biol. Bull. Woods Hole,* **127** : 108–118.

320. 1965. Respiration and osmotic behaviour of the copepod *Acartia tonsa* in diluted sea water. *Comp. Biochem. Physiol.* **14** : 155–165.

321. LARIMER, J. L. 1961. Measurement of ventilation volume in decapod crustaceans. *Physiol. Zool.* **34** : 158–166.

322. 1962. Responses of the crayfish heart during respiratory stress. *Physiol. Zool.* **35** : 179–186.

323. 1964. Sensory induced modifications of ventilation and heart rate in crayfish. *Comp. Biochem. Physiol.* **12** : 25–36.

324. 1964. The patterns of diffusion of oxygen across the crustacean (*Procambarus simulans*) gill membranes. *J. cell. comp. Physiol.* **64** : 139–148.

325. LARIMER, J. L. and GOLD, A. H. 1961. Responses of the crayfish *Procambarus simulans* to respiratory stress. *Physiol. Zool.* **34** : 167–176.

325b. LARIMER, J. L., TREVINO, D. L. and ASHBY, E. A. 1966. A comparison of caudal photoreceptors of epigeal and cavernicolous crayfish. *Comp. Biochem. Physiol.* **19** : 409–415.

326. LAVERACK, M. S. 1962. Responses of cuticular sense organs o[f] the lobster, *Homarus vulgaris* (Crustacea)—I, Hair peg organs as water current receptors. *Comp. Biochem. Physiol.* **5** : 319–325.

327. 1962. Responses of cuticular sense organs of the lobster, *Homarus vulgaris* (Crustacea)—II, Hair-fan organs as pressure receptors. *Comp. Biochem. Physiol.* **6** : 137–145.

328. 1963. Aspects of chemoreception in Crustacea. *Comp. Biochem. Physiol.* **8** : 141–151.

329. 1963. Responses of cuticular sense organs of the lobster, *Homarus vulgaris* (Crustacea)—III, Activity invoked in sense organs of the carapace. *Comp. Biochem. Physiol.* **10** : 261–272.

330. 1964. The antennular sense organs of *Panulirus argus*. *Comp. Biochem. Physiol.* **13** : 301–321.

331. LEAF, A. 1956. On the mechanism of fluid exchange of tissues *in vitro*. *Biochem. J.* **62** : 241–248.

332.   1959. Maintenance of concentration gradients and regulation of cell volume. *Ann. N.Y. Acad. Sci.* **72** : 396–404.

333. LEGRAND, J. J. and JUCHAULT, P. 1960. Disposition métamerique du tissu sécréteur de l'hormone mâle chez les different types d'Oniscoides. *C. R. Soc. Biol. Paris*, **250** : 754–756.

334. LIENEMANN, L. J. 1938. The green glands as a mechanism for osmotic and ionic regulation in the crayfish (*Cambarus clarkii*, Girard). *J. cell. comp. Physiol.* **11** : 149–161.

335. LOCKHEAD, J. H. and RESNER, R. 1959. Functions of the eyes and neurosecretion in Crustacea Anostraca. *Proc. 15th Internat. Congress Zool. Lond.* (1958) : 397–399.

336. LOCKWOOD, A. P. M. Unpublished observations.

337.   1959. The osmotic and ionic regulation of *Asellus aquaticus* (L). *J. exp. Biol.* **36** : 546–555.

338.   1960. Some effects of temperature and concentration of the medium on the ionic regulation of the isopod, *Asellus aquaticus* (L). *J. exp. Biol.* **37** : 614–630.

339.   1961. The urine of *Gammarus duebeni* and *G. pulex*. *J. exp. Biol.* **38** : 647–658.

340.   1962. The osmoregulation of Crustacea. *Biol. Rev.* **37** : 257–305.

341.   1959. The extra-haemolymph sodium of *Asellus aquaticus* (L). *J. exp. Biol.* **36** : 562–565.

342.   1964. The activation of the sodium uptake system at high blood concentrations in the amphipod, *Gammarus duebeni*. *J. exp. Biol.* **41** : 447–458.

343. LOCKWOOD, A. P. M. and CROGHAN, P. C. 1957. The chloride regulation of the brackish- and fresh-water races of *Mesidotea entomon* (L). *J. exp. Biol.* **34** : 253–258.

343b. LOCKWOOD, A. P. M. and CROGHAN, P. C. (In preparation.)

344. LOFTS, B. 1956. The effects of salinity changes on the respiratory rate of the prawn, *Palaemonetes varians* (Leach). *J. exp. Biol.* **33** : 730–736.

345. LOWENSTEIN, O. 1935. The respiratory rate of *Gammarus chevreuxi* in relation to differences in salinity. *J. exp. Biol.* **12** : 217–221.

346. DE LUCA, V. and PATANE, L. 1964. Further data on the determination of secondary sexual characters in Oniscoidea Isopoda (*Porcellio laevis*) (Translated). *Atti. Accad. Giogenia Sci. nat. Catania.* **13** : 1–19.

347. McWHINNIE, M. A. 1962. Gastrolith formation and calcium shift in the fresh water crayfish, *Orconectes virilis*. *Comp. Biochem. Physiol.* **7** : 1–14.

348. McWHINNIE, M. A. and CORKILL, A. J. 1964. The hexosemonophosphate pathway and its variation in the intermoult cycle in crayfish. *Comp. Biochem. Physiol.* **12** : 81–93.

349. McWHINNIE, M. A. and KIRCHENBERG, R. J. 1962. Crayfish hepatopancreas metabolism and the intermoult cycle. *Comp. Biochem. Physiol.* **6** : 117–128.

350. McWHINNIE, M. A. and Saller, P. N. 1960. Analysis of blood sugars in the crayfish, *Orconectes virilis*. *Comp. Biochem. Physiol.* **1** : 110–122.

351. McWHINNIE, M. A. and SCHEER, B. T. 1958. Blood glucose of the crab, *Hemigrapsus nudus*. *Science*, **128** : 90.

352. MADANMOHANRAO, G. and RAO, K. P. 1962. Oxygen consumption in a brackish water crustacean, *Sesarma plicatum* (Latreille) and a marine crustacean, *Lepas anserifera* L. *Crustaceana*, **4** : 75–81.

353. MALOEUF, N. S. R. 1937. Studies on the respiration and osmoregulation of animals. I, Aquatic animals without an oxygen transporter in their internal medium. *Z. vergl. Physiol.* **25** : 1–28.

354. —— 1938. On the kidney of the crayfish and on the uptake of chloride from fresh water by this animal. *Biol. Bull. Woods Hole*, **75** : 354–355.

355. —— 1938. Échanges d'eau et d'électrolytes chez un pagure. *P. longicarpus. Archs int. Physiol.* **47** : 1–23.

356. —— 1939. The blood of arthropods. *Q. Rev. Biol.* **14** : 149–191.

357. —— 1940. Secretion of inulin by the kidney of the crayfish. *Proc. Soc. exp. Biol. N.Y.* **45** : 873–875 ·

358. —— 1941. Experimental cytological evidence for an outward secretion of water by the nephric tubule of the crayfish. *Biol. Bull. Woods Hole*, **81**, 127–34 ; 235–260.

359. MARCHAL, P. 1892. Recherches anatomiques et physiologiques sur l'appareil excréteur des Crustacés decapodes. *Arch. zool. exp. gen.* **10** : 57–275.

360. MARSHALL, S. M. and ORR, A. P. 1952. On the biology of *Calanus finmarchicus*. VII, Factors affecting egg production. *J. mar. biol. Ass. U.K.* **30** : 527–547.

361. —— 1955. Experimental feeding of the copepod *Calanus finmarchicus* (Gunner) on phytoplankton cultures labelled with radio-active carbon (^{14}C). *Deep Sea Res. Suppl.* **3** : 110–114.

362. MARSLAND, D. A. 1944. Mechanism of pigment displacement in unicellular chromatophores. *Biol. Bull. Woods Hole*, **87**, 252–261.

363. MARTIN, A. W. 1957. Recent advances in knowledge of invertebrate renal function. In : *Recent Advances in Invertebrate Physiology*. Univ. Oregon Press.

364.   1958. Comparative physiology (excretion). *A. rev. Physiol.* **20** : 225–421.

365. MARTIN, A. W., HARRISON, F. M., HUSTON, M. J. and STEWART, D. M. 1958. The blood volumes of some representative molluscs. *J. exp. Biol.* **35** : 260–279.

365b. MATTHEWS, T. H. (Personal communication.)

366. MAUCHLINE, J. 1959. The biology of the Euphausid crustacean *Meganyctiphanes norvegica* (M. Sars). *Proc. Roy. Soc. Edinb. B.* **67** : 141–179.

367. MAYNARD, D. M. 1953. Integration in the cardiac ganglion of *Homarus*. *Biol. Bull. Woods Hole*, **105** : 367.

368.   1960. Circulation and heart function. In : *The Physiology of the Crustacea*, ed. Waterman, T. H. Academic Press.

369.   1961. Thoracic neurosecretory structures in Brachyura. II, Secretory neurones. *Gen. & comp. Endocr.* **1** : 237–263.

370.   1961. Thoracic neurosecretory structures in Brachyura. I, Gross anatomy. *Biol. Bull. Woods Hole*, **121** : 316–329.

371. MAYNARD, D. M. and MAYNARD, E. A. 1962. Thoracic neurosecretory structures in Brachyura. III, Microanatomy of peripheral structures. *Gen. & comp. Endocr.* **2** : 12–18.

372. MAYNARD, D. M. and WELSH, J. H. 1959. Neurohormones of the pericardial organs of Brachyuran Crustacea. *J. Physiol.* **149** : 215–227.

373. MEENAKSHI, V. R. and SCHEER, B. T. 1958. Acid mucopolysaccharide of the crustacean cuticle. *Science* **130** : 1189–1190.

374.   1961. Metabolism of glucose in the crabs *Cancer magister* and *Hemigrapsus nudus*. *Comp. Biochem. Physiol.* **3** : 30–41.

375. MELLON, D. 1963. Electrical response from dually innervated tactile receptors on the thorax of the crayfish. *J. exp. Biol.* **40** : 137–148.

376. MILLIKAN, G. A. 1933. The kinetics of blood pigments : Haemocyanin and haemoglobin. *J. Physiol.* **79** : 158–179.

377. MÜLLER, I. 1943. Untersuchungen zur Gesetzlichkeit des Wachstums. X Weiteres zur Frage der Abhängigkeit der Atmung von Körpergrosse. *Biol. Zbl.* **63** : 446–453.

378. MUNDAY, K. A. and THOMPSON, B. D. 1962. The preparation and properties of sub-cellular respiring particles (mitochondria) from the hepatopancreas of *Carcinus maenas*. *Comp. Biochem. Physiol.* **5** : 95–112.

379. MUNTWYLER, E., GRIFFIN, G. E. and ARENDS, R. L. 1953. Muscle electrolyte composition and balances of nitrogen and potassium in potassium deficient rats. *Am. J. Physiol.* **174** : 283–288.

380. NAGANO, T. 1958. Physiological studies on the pigmentary system of crustacea. XIII, Intake and output of water regulated by the eyestalk hormone of a crayfish. *Sci. Rep. Tohoku Univ.* **24** : 9–13.

381. NAGEL, H. 1934. Die Aufgaben der Excretionsorgane und der Kiemen bei der Osmoregulation von *Carcinus maenas*. *Z. verg. Physiol.* **21** : 468–491.

382. NEEDHAM, A. E. 1957. Factors affecting nitrogen excretion in *Carcinides maenas*. *Physiol. comp. Oecol.* **4** : 209–239.

383. NEEDHAM, J. 1932. The energy sources in ontogeny, VII. *J. exp. Biol.* **10** : 79–87.

384. NEILAND, K. A. and SCHEER, B. T. 1954. The influence of fasting and of sinus gland removal on body composition of *Hemigrapsus nudus* (Part V of the hormonal regulation of metabolism in crustacea). *Physiologia compar. et Oecol.* **3** : 321–326.

385. NICHOLLS, A. G. 1931. Studies on *Ligia oceanica*. Part II, The processes of feeding, digestion and absorption, with a description of the structure of the foregut. *J. mar. biol. Ass. U.K.* **17** : 675–706.

386. ———— 1946. Syncarida in relation to the interstitial habitat. *Nature,* **158** : 934.

387. NUMANOI, H. 1939. Behaviour of blood calcium in the formation of gastroliths in some decapod crustaceans. *Jap. J. Zool.* **8** : 357–363.

388. OLMSTEAD, J. M. D. and BAUMBERGER, J. P. 1923. Form and growth of grapsoid crabs. *J. Morph.* **18** : 279.

389. ORKAND, P. K. 1962. Chemical inhibition of contraction in directly stimulated crayfish muscle fibres. *J. Physiol.* **164** : 103–115.

390. ———— 1962. The relation between membrane potential and contraction in single crayfish muscle fibres. *J. Physiol.* **161** : 143–159.

391. PANIKKAR, N. K. 1941. Osmoregulation in some Palaemonid prawns. *J. mar. biol. Ass. U.K.* **25** : 317–359.

392. ———— 1941. Osmotic behaviour of the fairy shrimp *Chirocephalus diaphanus*. *J. exp. Biol.* **18** : 110–114.

393. PANIKKAR, N. K. 1950. Physiological aspects of adaptation to estuarine conditions. *Proc. Indo-Pacif. Fish Council*, **2** : 168–175.

394. PANOUSE, J. B. 1943. Influence de l'ablation du pédoncle oculaire sur la croissance de l'ovaire chez la crevette *Leander serratus*. *C. R. Acad. Sci. Paris*, **217** : 553–555.

395. ———— 1944. L'action de la glande du sinus sur l'ovaire chez la crevette, *Leander serratus*. *C. R. Acad Sci. Paris*, **218** : 293–294.

396. ———— 1946. Recherches sur les phénomènes humoraux chez les Crustacés. *Ann. Inst. Océanogr.* **23** : 65–147.

397. PANTIN, C. F. A. 1931. The origin of the composition of the body fluids of animals. *Biol. Rev.* **6** : 459–482.

398. PAPI, F. 1960. Orientation by night : the moon. *Cold Spring Har. Symp.* **25** : 475–480.

399. PARDI, L. 1960. Innate components in the solar orientation of littoral amphipods. *Cold Spring Har. Symp.* **25** : 395–401.

400. PARK, T. 1946. A further report on tolerance experiments by ecological classes. *Ecology*, **26** : 305–308.

401. PARK, T., GREGG, R. E. and LUTTERMAN, C. Z. 1940. Toleration experiments by ecology classes. *Ecology*, **21** : 109–111.

402. PARRY, G. 1953. Osmotic and ionic regulation in the isopod crustacean, *Ligia oceanica*. *J. exp. Biol.* **30** : 567–574.

403. ———— 1954. Ionic regulation in the Palaemonid prawn, *Palaemon serratus*. *J. exp. Biol.* **30** : 601–613.

404. ———— 1955. Urine production by the antennal glands of *Palaemonetes varians* (Leach). *J. exp. Biol.* **32** : 408–422.

405. ———— 1957. Osmoregulation in some fresh-water prawns. *J. exp. Biol.* **34** : 417–423.

406. PASSANO, L. M. 1953. Neurosecretory control of moulting in crabs by the X organ—sinus gland complex. *Physiologia comp. et Oecol.* **3** : 155–189.

407. ———— 1954. Phase microscopic observations of the neurosecretory product of the crustacean X organ. *Pubbl. Staz. zool. Napoli*, **24** : 72–73.

408. PASSANO, L. M. and JYSSUM, S. 1963. The role of the Y-organ in crab proecdysis and limb regeneration. *Comp. Biochem. Physiol.* **9** : 195–214.

409. PEARSE quoted in 220.

410. PEREZ, C. and BLOCK-RAPHAEL. 1946. Note préliminaire sur la présence d'un pigment respiratoire chez le *Septosaccus Cuenoti* (Dubosq.). *C. R. Acad. Sci. Paris*, **223** : 840.

411. PESCH, L. A. and TOPPER, Y. J. 1964. The liver and carbohydrate metabolism. In : *The Liver*, ed. Rouiller, Ch. IV. Academic Press.

412. PETERS, H. 1935. Über den Einfluss des Salzgehaltes in Aussenmedium auf den Bau und die Funktion der Excretionsorgane Decapoda Crustaceen. *Z. Morph. Ökol. Tiere*, **30** : 355–381.

413. PHILLIPS, J. E. 1964. Rectal absorption in the desert locust, *Schistocerca gregaria*. Forskal. I, Water. *J. exp. Biol.* **41** : 15–38.

414. PICKEN, L. E. R. 1936. The mechanism of urine formation in invertebrates. I, The excretion mechanisms in certain Arthropods. *J. exp. Biol.* **13** : 309–328.

415. PINHEY, K. G. 1930. Tyrosinase in crustacean blood. *J. exp. Biol.* **7** : 19–36.

416. PILGRIM, R. L. C. 1960. Muscle receptor organs in some decapod Crustacea. *Comp. Biochem. Physiol.* **1** : 248–257.

416b. PILGRIM, R. L. C. 1964. Stretch receptor organs in *Squilla mantis* Latr. (Crustacea Stomatopoda). *J. exp. Biol.* **41** : 793–804.

417. PORTER, K. R. and PALADE, G. E. 1957. Studies on the endoplasmic reticulum, III. *J. biophys. Biochim. Cytol.* **3** : 269–300.

418. POTTER, D. D. 1954. Histology of the neurosecretory system of the blue crab, *Callinectes sapidus*. *Anat. Rec.* **120** : 716.

419. POTTS, W. T. W. 1959. The sodium fluxes in the muscle fibres of a marine and a fresh-water lamellibranch. *J. exp. Biol.* **36** : 676–689.

420. POWELL, B. L. 1962. The responses of the chromatophores of *Carcinus maenas* to light and temperature. *Crustaceana*, **4** : 93–102.

421. PRESTON, J. B. and KENNEDY, D. 1960. Integratory synaptic mechanisms in the caudal ganglion of the crayfish. *J. gen. Physiol.* **43** : 671–681.

421b. PRESTON, J. B. and KENNEDY, D. 1962. Spontaneous activity in Crustacean neurones. *J. gen. Physiol.* **45** : 821–836.

422. PROSSER, C. L. 1934. Action potentials in the nervous system of the crayfish. II, Responses to illumination of the eye and caudal ganglion. *J. cell. comp. Physiol.* **4** : 363–377.

423. —— 1935. Action potentials in the nervous system of the crayfish. III, Central responses to proprioceptive and tactile stimulation. *J. comp. Neurol.* **62** : 495–505.

424. —— 1955. Physiological variation in animals. *Biol. Rev.* **30** : 229–262.

425. PROSSER, C. L. and WEINSTEIN, S. J. F. 1950. Comparison of blood volume in animals with open and with closed circulatory systems. *Physiol. Zool.* **23** : 113–124.

426. PROSSER, C. L., BISHOP, D. W., BROWN, F. A., JAHN, T. L. and WULFF, V. J. 1950. *Comparative Animal Physiology.* W. B. Saunders Co.

427. PROSSER, C. L., GREEN, S. W. and CHOW, T. S. 1955. Ionic and osmotic concentrations in the blood and urine of *Pachygrapsus crassipes* acclimated to different salinities. *Biol. Bull. Woods Hole,* **109** : 99–107.

427b. QUIN, D. J. and LANE, C. E. 1966. Ionic regulation and Na$^+$–K$^+$ stimulated A.T.P.ase activity in the land crab *Cardiosoma guanhami. Comp. Biochem. Physiol.* **19** : 533–544.

428. RALPH, R. 1965. Thesis, Southampton.

429. RAMSAY, J. A. 1961. The comparative physiology of renal function of invertebrates. In : *The Cell and the Organism.* Cambridge Univ. Press.

430. RAY, D. L. and JULIAN, J. R. 1952. Occurrence of cellulase in *Limnoria. Nature,* **169** : 32–33.

431. RAYMONT, J. E. G. 1964. *Plankton and Productivity.* Pergamon Press.

432. RAYMONT, J. E. G. and CONOVER, R. J. 1961. Further investigation on the carbohydrate content of marine zooplankton. *Limnol. et Oceanogr.* **6** : 154–164.

433. RAYMONT, J. E. G. and KRISHNASWAMY, S. 1960. Carbohydrates in some marine planktonic animals. *J. mar. biol. Ass. U.K.* **39** : 239–248.

434. RAYMONT, J. E. G., AUSTIN, J. and LINFORD, E. 1964. Biochemical studies on marine zooplankton. I, The biochemical composition of *Neomysis integer. J. Conseil.* **28** : 354–363.

435. REDFIELD, A. C. 1934. The haemocyanins. *Biol. Rev.* **9** : 175–212.

436. REDFIELD, A. C., COOLIDGE, T. and HURD, A. L. 1926. The transport of oxygen and carbon dioxide by some bloods containing haemocyanin. *J. Biol. Chem.* **69** : 475–509.

437. REDMOND, J. R. 1955. The respiratory function of haemocyanin in crustacea. *J. cell. comp. Physiol.* **46** : 209–247.

438. REGNARD, P. and BLANCHARD, B. 1883. Note sur la présence de l'hémoglobin dans le sang des Crustacés branchiopodes. *Zool. Anz.* **6** : 253.

439. RENAUD, L. 1949. Le cycle des réserves organiques chez les crustacés décapodes. *Ann. Inst. Océanogr., Monaco,* **24** : 259–357.

440. REUBEN, J. P., GIRARDIER, L. and GRUNDFEST, H. 1964. Water transfer and cell structure in isolated crayfish muscle fibres. *J. gen. Physiol.* **47** : 1141–1194.

441. RICE, A. L. 1961. The responses of certain mysids to changes in hydrostatic pressure. *J. exp. Biol.* **38** : 391–402.

442. RICE, A. L. 1964. Observations on the effects of changes in pressure on the behaviour of some marine animals. *J. mar. biol. Ass. U.K.* **44** : 163–175.

443. RIEGEL, J. A. 1959. Some aspects of osmoregulation in two species of sphaeromid isopod crustacea. *Biol. Bull. Woods Hole,* **116** : 272–284.

444. ——— 1961. The influence of water loading and low temperature on certain functional aspects of the crayfish antennal gland. *J. exp. Biol.* **38** : 291–300.

445. ——— 1963. Micropuncture studies of chloride concentration and osmotic pressure in the crayfish antennal gland. *J. exp. Biol.* **40** : 487–492.

446. ——— 1965. Micropuncture studies of the concentration of sodium potassium and inulin in the crayfish antennal gland. *J. exp. Biol.* **42** : 379–384.

446b. ——— 1966. Micropuncture studies of formed-body secretions by the excretory organs of the crayfish, frog and stick insect. *J. exp. Biol.* **44** : 379–385.

446c. ——— 1966. Analysis of formed bodies in urine removed from the crayfish antennal gland by micropuncture. *J. exp. Biol.* **44** : 387–396.

447. Quoted by 449.

448. RIEGEL, J. A. and KIRSCHNER, L. B. 1960. The excretion of inulin and glucose by the crayfish antennal gland. *Biol. Bull. Woods Hole,* **118** : 296–307.

449. RIEGEL, J. A. and LOCKWOOD, A. P. M. 1961. The role of the antennal gland in the osmotic and ionic regulation of *Carcinus maenas*. *J. exp. Biol.* **38** : 491–499.

450. RIEGEL, J. A. and PARKER, R. A. 1960. A comparative study of crayfish blood volumes. *Comp. Biochem. Physiol.* **1** : 302–304.

451. ROBERTS, T. W. 1944. Light, eyestalk chemical and certain factors as regulators of community activity for the crayfish, *Cambarus virilis* (Hagen). *Ecol. Monographs,* **14** : 359–392.

452. ROBERTS, J. 1957. Thermal acclimation of metabolism in the crab, *Pachygrapsus crassipes* Randall. I, The influence of body size, starvation and moulting. *Physiol. Zool.* **30** : 232–242.

453. ——— 1959. Thermal acclimation of metabolism in the crab, *Pachygrapsus crassipes* Randall. II, Mechanisms and the influence of season and latitude. *Physiol. Zool.* **30** : 243–255.

454. ROBERTSON, J. D. 1939. The inorganic composition of body fluids of three marine invertebrates. *J. exp. Biol.* **16** : 387.

454a. 1949. Ionic regulation in some marine invertebrates. *J. exp. Biol.* **26** : 182–200.

455. 1953. Further studies on ionic regulation in marine invertebrates. *J. exp. Biol.* **30** : 277–296.

456. 1960. Ionic regulation in the crab *Carcinus maenas* (L) in relation to the moulting cycle. *Comp. Biochem. Physiol.* **1** : 183–212.

457. 1961. Studies on the chemical composition of muscle tissue. II, The abdominal flexor muscles of the lobster *Nephrops norvegicus.* (L). *J. exp. Biol.* **38** : 707–728.

458. RUCK, P. and JAHN, T. L. 1954. Electrical studies on the compound eye of *Ligia occidentalis* (Dana). *J. gen. Physiol.* **37** : 825–849.

459. SCHALLEK, W. 1945. Action of potassium on bound acetylcholine in lobster nerve cord. *J. cell. comp. Physiol.* **26** : 15–24.

460. SCHEER, B. T. 1959. The hormonal control of metabolism in crustaceans. IX, Carbohydrate metabolism in the transition from intermoult to premoult in *Carcinides maenas. Biol. Bull. Woods Hole,* **116** : 175–183.

461. 1960. Aspects of the intermoult cycle in Natantians. *Comp. Biochem. Physiol.* **1** : 3–18.

462. SCHEER, B. T. and SCHEER, M. A. R. 1951. Blood sugar in spiny lobsters. Part I of the hormonal regulation of metabolism in Crustaceans. *Physiologia compar. et Oecol.* **2** : 198–209.

463. 1954. The hormonal control of metabolism in crustaceans. VII, Moulting and colour change in the prawn *Leander serratus. Publ. Staz. zool. Napoli,* **25** : 397–418.

464. SCHEER, B. T., SCHWABE, C. W. and SCHEER, M. A. 1952. Tissue oxidations in crustaceans. III, On hormonal regulation of metabolism in crustaceans. *Physiologia compar. et Oecol.* **2** : 327–338.

465. SCHLIEPER, C. 1929. Einwirkung niederer Salzkonzentrationem auf marine Organismen. *Z. vergl. Physiol.* **9** : 478.

466. SCHOFFENIELS, E. 1960. Origine des acides aminés intervenant dans la régulation de la pression osmotique intracellulaire de *Eriocheir sinensis*, Milne Edwards. *Archs int. Physiol. et Biochem.* **68** : 696–7.

466b. 1965. L-Glutamic acid dehydrogenase activity in the gills of *Palinurus vulgaris* Latr. *Archs int. Physiol. Biochem.* **73** : 73–80.

467. SCHOLANDER, P. F., FLAGG, W., WALTERS, V. and IRVING, L. 1953. Climatic adaptation in arctic and tropical poikilotherms. *Physiol. Zool.* **26** : 67–92.

468. SCHOLLES, W. 1933. Über die Mineralregulation wasserlebender Evertebraten. *Z. vergl. Physiol.* **19** : 522–554.

469. SCHÖNE, H. 1951. Die statische Gleichgewichalsorientierung dekapoder Crustaceen. *Ver. deut. zool. Ges.* **16** : 157–162. (Quoted from Cohen).

470. SCHÖNE, H. and SCHÖNE, H. 1961. Eyestalk movements induced by polarised light in the ghost crab, *Ocypode quadrata*. *Science*, **134** : 675–676.

471. SCHWABE, E. 1933. Über die Osmoregulation verschiedener Krebse (Malacostracen). *Z. vergl. Physiol.* **19** : 183–236.

472. SCHWARTZKOPFF, J. 1955. Vergleichende Untersuchungen der Hertzfrequenz bei Krebsen. *Biol. Zentralb.* **74** : 480–497.

473. SCUDAMORE, H. H. 1947. Influence of sinus glands upon moulting and associated changes in crayfish. *Physiol. Zool.* **20** : 187–208.

474. SEEGAR, J. 1934. Die Atmungsbewegungen von *Astacus fluviatilis Z. vergl. Physiol.* **21** : 492–515.

475. SHAW, J. 1955. Ionic regulation in the muscle fibres of *Carcinus maenas*. I, The electrolyte composition of single fibres. *J. exp. Biol.* **32** : 383–396.

476.   1955. Ionic regulation in the muscle fibres of *Carcinus maenas*. II The effect of reduced blood concentration. *J. exp. Biol.* **32** : 664–680.

477.   1958. Further studies on ionic regulation in the muscle fibres of *Carcinus maenas*. *J. exp. Biol.* **35** : 902–919.

478.   1958. Osmoregulation in the muscle fibres of *Carcinus maenas*. *J. exp. Biol.* **35** : 920–929.

479.   1959. Salt and water balance in the East African fresh-water crab, *Potamon niloticus* (M. Edw.) *J. exp. Biol.* **36** : 157–176.

480.   (Same as 479.)

481.   1959. Solute and water balance in the muscle fibres of the East African fresh-water crab. *Potamon niloticus* (M. Edw.). *J. exp. Biol.* 36 : 145–156.

482.   1959. The absorption of sodium ions by the crayfish *Astacus pallipes*, Lereboullet. I, The effect of external and internal sodium concentrations. *J. exp. Biol.* **36** : 126–144.

483.   1960. The absorption of chloride ions by the crayfish *Astacus pallipes*. *J. exp. Biol.* **37** : 557–572.

484. SHAW, J. 1960. The absorption of sodium ions by the crayfish *Astacus pallipes* Lereboullet, II. *J. exp. Biol.* **37** : 534–547.

485.   1960. The absorption of sodium ions by the crayfish *Astacus pallipes* Lereboullet. III, The effect of other cations in the external medium. *J. exp. Biol.* **37** : 548–556.

486.   1961. Sodium balance in *Eriocheir sinensis* (M. Edw.) The adaptation of crustacea to fresh water. *J. exp. Biol.* **38** : 153–162.

487.   1961. Studies on ionic regulation in *Carcinus maenas* (L). *J. exp. Biol.* **38** : 135–152.

488. SHIRAISHI, K. and PROVASOLI, L. 1959. Growth factors as supplements to inadequate algal food for *Tigriopus japonicus*. *Int. Ocean. Congress Preprints. A.A.A.S., Washington* : 951–952.

489. SMITH, R. I. 1939. Acetylcholine in the nervous tissue and blood of crayfish. *J. cell. comp. Physiol.* **13** : 335 – 344.

490. SPENCER, J. O. and EDNEY, E. B. 1954. The absorption of water by woodlice. *J. exp. Biol.* **31** : 491–496.

491.   1955. Cutaneous respiration in woodlice. *J. exp. Biol.* **32** : 256–269.

492. SPROSTON, N. G. and HARTLEY, P. H. T. 1942. Observations on bionomics and physiology of *Trebius caudatus* and *Lernaeocera branchialis* Copepoda. *J. mar. biol. Ass. U.K.* **25** : 393–417.

493. STEDMAN, E. and STEDMAN, E. 1925. Haemocyanin. Part I, The dissociation curves of the oxyhaemocyanin in the blood of some decapod crustaceans. *Biochem. J.* **19** : 544–551.

494. STEHR, W. C. 1931. The activating influence of light upon certain aquatic arthropods. *J. exp. Zool.* **59** : 297–335.

495. STEPHENS, G. C. 1955. Induction of moulting in the crayfish, *Cambarus* by modification of daily photoperiod. *Biol. Bull. Woods Hole.* **108** : 235–241.

496. STONE, D. and SHAPIRO, S. 1948. Investigations of free and bound potassium in rat brain and muscle. *Am. J. Physiol.* **155** : 141–146.

497. STOTT, F. C. 1932. Einige vorlaufige Versucle über Veränderungen des Blutguchers bei Decapoden. *Biochem. Z.* **248** : 55–64.

498a. SUGI, H. and KOSAKA, K. 1964. Summation of contraction in single crayfish muscle fibres. *Jap. J. Physiol.* **14** : 450–467.

498b. SVEDBERG, T. 1939. A discussion on the protein molecule. *Proc. R. Soc. B.,* **127** : 1–17.

499. SWEDMARK, B. 1964. The interstitial fauna of marine sand. *Biol. Rev.* **39** : 1–42.

500. TAKEDA, K. and KENNEDY, D. 1965. The mechanism of discharge pattern formation in crayfish interneurones. *J. gen. Physiol.* **48** 435–453.

500b. TACHEUCHI, A. and TAKEUCHI, N. 1964. The effects on crayfish muscle of iontophoretically applied glutamate. *J. Physiol.* **170** : 269–317.

500c. ———— 1966. On the permeability of the presynaptic terminal of the crayfish neuromuscular junction during synaptic inhibition and the action of γ-amino butyric acid. *J. Physiol.* **183** : 433–449.

501. TCHERNIGOUTZEFF, C. 1959. *C. R. Acad. Sci. Paris*, **248** : 600–606.

502. TEAL, J. M. 1958. Distribution of fiddler crabs in Georgia salt marsh. *Ecology*, **39** : 185–193.

503. ———— 1959. Respiration of crabs in Georgia salt marshes and its relation to their ecology. *Physiol. Zool.* **32** : 1–14.

504. TEYEN, F. J., SUDAK, F. N. and CLAFF, C. L. 1959. Oxygen consumption in *Uca pugnax* and *U. pugilator* before and after eyestalk removal. *Biol. Bull. Woods Hole*, **117** : 429.

505. TOPPEL, A. L. 1960. Cytochromes of muscles of marine invertebrates. *J. cell. comp. Physiol.* **55** : 111–126.

506. TAYLOR, A. W. 1960. Thesis. Southampton.

507. THOMAS, A. J. 1954. The oxygen uptake of the lobster (Homarus vulgaris EdW.). *J. exp. Biol.* **31** : 228–251.

508. TRAUTWEIN, W., KUFFLER, S. W. and EDWARDS, C. 1956. Changes in membrane characteristics of heart muscle during inhibition. *J. gen. Physiol.* **40** : 135–145.

509. TRAVIS, D. F. 1954. The moulting cycle of the spiny lobster, *Panulirus argus*, Latreille. I, Moulting and growth in laboratory maintained individuals. *Biol. Bull. Woods Hole*, **107** : 433–450.

510. ———— 1955. The moulting cycle of the spiny lobster, *Panulirus argus*, Latreille. II, Pre-ecdysial histological and histochemical changes in the hepatopancreas and integumental tissues. *Biol. Bull. Woods Hole*, **108** : 88–112.

511. ———— 1957. Moulting cycle in the spiny lobster, *Panulirus argus*, Latreille. IV, Post-ecdysial histological and histochemical changes in the hepatopancreas and integumental tissues. *Biol. Bull. Woods Hole*, **113** : 451–479.

512. TREHERNE, J. E. 1962. The physiology of absorption from the alimentary canal in insects. *Viewpoints in Biology*, **1** : 201–241.

513. USSING, H. H. Quoted by Krogh.

514. VAN DER KLOOT, W. G. 1960. Picrotoxin and the inhibitory system of crayfish muscle. In : *Inhibition in the Nervous System and γ-aminobutyric Acid* (pp. 409–412). Pergamon.

515. VAN HARREVALD, A. 1939. *J. comp. Neurol.* **70** : 267–284.

516. VAN HARREVALD, A. and WIERSMA, C. A. G. 1937. The triple innervation of crayfish muscle and its function in contraction and inhibition. *J. exp. Biol.* **14** : 448–464.

516b. VAN WEEL, P. B. and CHRISTOFFERSON, J. P. 1966. Electrophysiological studies on perception in the antennulae of certain crabs. *Physiol. Zool.* **39** : 317–325.

517. VAN WEEL, P. B., RANDALL, J. E. and TAKATA, M. 1954. Observations on the oxygen consumption of certain marine crustaceans. *Pacif. Sci.* **8** : 209–218.

518. VERHOEF, K. W. 1917. Zur Kentniss der Atmung und der Atmungsorgane der Isopoda Oniscoidea. *Biol. Zbl.* **37** : 113–127.

519. VERNBERG, F. J. 1956. Study of the oxygen consumption of excised tissues of certain marine decapod crustacea in relation to habitat. *Physiol. Zool.* **29** : 227–234.

520. —— 1959. Studies on the physiological variation between tropical and temperate zone fiddler crabs of the genus *Uca*. II, Oxygen consumption of whole organisms. *Biol. Bull. Woods Hole*, **117** : 163–184.

521. —— 1959. Studies on the physiological variation between tropical and temperate zone fiddler crabs of the genus *Uca*. III, The influence of temperature acclimation on oxygen consumption of whole organisms. *Biol. Bull. Woods Hole*, **117** : 582–593.

522. —— 1962. Comparative physiology. Latitudinal effects on physiological properties of animal populations. *A. Rev. Physiol.* **24** : 517–546.

522b. VERNBERG, F. J. and VERNBERG, W. B. 1966. Studies on the physiological variation between tropical and temperate zone fiddler crabs of the genus *Uca*. VII, Metabolic-temperature acclimatisation responses in southern hemisphere crabs. *Comp. Biochem. Physiol.* **19** : 489–524.

523. VERNE, J. 1924. Note histochimique sur le métabolisme du glycogène pendant la mue chez les crustacés. *C. R. Soc. Biol. Paris*, **90** : 186–188.

524. VERNET-CORNUBERT, G. and DEMEUSY, N. 1955. Influence de l'ablation des pédoncles oculaires sur les charactères sexuels externes des femelles pubères de *Carcinus maenas* et de *Pachygrapsus marmoratus*. *C. R. Acad. Sci.* **240** : 360–361.

525. VONK, H. J. 1947. La présence d'acides bilaires et la résorption des acides gras chez les invertébrés. *Bulles. soc. Chim. Biol.* **29** : 94–96.

526. —— 1955. Comparative physiology : nutrition, feeding and digestion. *A. Rev. Physiol.* **17** : 483–498.

527. —— 1960. Digestion and metabolism. In : *The Physiology of Crustacea*. ed. Waterman, T. H. Academic Press.

528. WALD, G., BROWN, P. K. and GIBBONS, I. R. 1962. Visual excitation : a chemo-anatomical study. *Symp. Soc. Exp. Biol.* **16** : 32–57.

529. WALD, G. and HUBBARD, R. 1957. Visual pigment of a decapod crustacean, the lobster. *Nature*, **180** : 278–280.

530. WALLOP, J. N. and BOOT, L. M. 1950. Studies on cholinesterase in *Carcinus maenas*. *Biochim. biophys. Acta*, **4** : 566–571.

531. WANG, D. H. and SCHEER, B. T. 1962. Effect of eyestalk extract on U.D.P.G.—glycogen transglucosylase in crab muscle. *Life Sciences*, **1** : 209–211.

532. WANG, D. H. and SCHEER, B. T. 1963. U.D.P.G.—glycogen transglucosylase and a natural inhibitor in crustacean tissue. *Comp. Biochem. Physiol.* **9** : 263–274.

533. WALSHE-MAETZ, B. M. 1956. Contrôle respiratoire et métabolisme chez les crustacés. *Vie et Milieu*, **7** : 523–543.

534. WATANABE, A. and GRUNDFEST, H. 1961. Impulse propagation at the septal and commissural junctions of crayfish lateral giant axons. *J. gen. Physiol.* **45** : 267–308.

535. WATERMAN, T. H. 1941. A comparative study of the effects of ions on the whole nerve and isolated single nerve fibre preparations of crustacean neuromuscular systems. *J. cell. comp. Physiol.* **18** : 109–126.

536. —— 1960. Light sensitivity and vision. In : *The Physiology of Crustacea.* ed. Waterman, T. H. Academic Press.

537. WATERMAN, T. H. and WIERSMA, C. A. G. 1963. Electrical responses in decapod crustacea visual systems. *J. cell. comp. Physiol.* **61** : 1–16.

538. WATERMAN, T. H., WIERSMA, C. A. G. and BUSH, B. H. M. 1964. Afferent visual responses in the optic nerve of the crab, *Podophthalmus*. *J. cell. comp. Physiol.* **63** : 135–155.

539. WEBB, D. A. 1940. Ionic regulation in *Carcinus maenas*. *Proc. R. Soc.*, *B.* **129** : 107–136.

540. WEBB, H. M. and BROWN, F. A. 1958. The repetition of pattern in the respiration of *Uca pugnax*. *Biol. Bull. Woods Hole*, **115** : 303–318.

541. WEBB, H. M. and BROWN, F. A. 1961. Seasonal variations in O_2 consumption of *Uca pugnax*. *Biol. Bull Woods Hole*, **121** : 561–571.

542. WEBER, quoted by Jenkin.

543. WELSH, J. H. 1934. The caudal photoreceptor and response of the crayfish to light. *J. cell. comp. Physiol.* **4** : 379–388.

544. —— (Same as 543.)

545. WELSH, J. H.   1939.   The action of eyestalk extracts on retinal pigment migration in the crayfish *Cambarus bartoni*.   *Biol. Bull. Woods Hole*, **77** :   119–125.

546. WELSH, J. H. and MOORHEAD, M.   1960.   The quantitative distribution of 5-hydroxytryptamine in the invertebrates, especially in their nervous systems.   *J. Neurochem.* **6** :   146–169.

547. WERMAN, R. and GRUNDFEST, H.   1961.   Graded and all-or-none electrogenesis in arthropod muscle.   II, The effects of alkaline earth and onium ions on lobster muscle fibres.   *J. gen. Physiol.* **44** :   997.

548. WERNTZ, H. O.   1963.   Osmotic regulation in marine and brackish water gammarids (Amphipoda).   *Biol. Bull. Woods Hole*, **124** :   225–239.

549. WEYMOUTH, F. W., CRISMAN, J. M., HALL, V. E., BELDING, H. S. and FIELD J.   1944.   II, Total and tissue respiration in relation to body weight.   A comparison of the kelp crab with other crustaceans and with mammals.   *Physiol. Zool.* **17** :   50–71.

550. WHITEAR, M.   1960.   Chordotonal organs in crustacea.   *Nature*, **187** :   522–523.

550a.   1962.   The fine structure of crustacean proprioceptors, I, The chordotonal organs in the legs of the shore crab *Carcinus maenas*.   *Phil. Trans. R. Soc., B.* **245** :   291–324.

551. WIENS, A. W. and ARMITAGE, K. B.   1961.   The Oxygen consumption of the crayfish *Orconectes immunis* and *Orconectes nais* in response to temperature and to Oxygen saturation.   *Physiol. Zool.* **34** :   39–54.

552. WIERSMA, C. A. G.   1947.   Giant nerve fibre system of the crayfish.   A contribution to comparative physiology of the synapse.   *J. Neurophysiol.* **10** :   23–38.

553.   1959.   Movement receptors in decapod crustacea.   *J. mar. biol. Ass. U.K.* **38** :   143–152.

554.   1960.   The Neuro-muscular system.   In :   *The Physiology of Crustacea*, Vol. II, pp. 191–240, ed. Waterman, T. H.

555. WIERSMA, C. A. G. and BOETTIGER, E. G.   1959.   Undirectional movement fibres from a proprioceptive organ of the crab *Carcinus maenas*.   *J. exp. Biol.* **36** :   102–112.

556. WIERSMA, C. A. G. and HUGHES, G. M.   1961.   On the functional anatomy of neuronal units in the abdominal cord of the crayfish.   *Procambarus clarkii* (Girard).   *J. comp. Neurol.* **116** :   209–228.

557. WIERSMA, C. A. G. and RIPLEY, S. H.   1952.   Innervation patterns of crustacean limbs.   *Physiologia compar. et Oecol.* **2** : 391–405.

558. WIERSMA, C. A. G. and RIPLEY, S. H. 1954. Further functional differences between fast and slow contractions in certain crustacean muscles. *Physiologia compar. et Oecol.* **3** : 327–336.

559. WIERSMA, C. A. G. and VAN HARREVALD, A. 1938. The influence of the frequency of stimulation on the slow and fast contraction in crustacean muscle. *Physiol. Zool.* **11** : 75–81.

559b. WIERSMA, C. A. G. and PILGRIM, R. L. C. 1961. Thoracic stretch receptors in crayfish and rock-lobster. *Comp. Biochem. Physiol.* **2** : 51–64.

560. WIERSMA, C. A. G., FURSCHPAN, E. and FLOREY, E. 1953. Physiological and pharmacological observations on muscle receptor organs of the crayfish, *Cambarus clarkii* Girard. *J. exp. Biol.* **30** : 136–150.

561. WIRZ, H. 1961. Kidney, water and electrolyte metabolism. *A. Rev. Physiol.* **23** : 577–606.

561b. WYSE, G. A. and MAYNARD, D. M. 1965. Joint receptors in the antennule of *Panulirus argus*. *J. exp. Biol.* **42** : 521–536.

561c. YALDWYN, J. C. 1966. Protandrous hermaphroditism in decapod prawns of the families Hippolytidae and Compylonolidae. *Nature*, **209** : 1366.

562. YOKOE, Y. and YASUMASU, I. 1964. The distribution of cellulase in invertebrates. *Comp. Biochem. Physiol.* **13** : 323–338.

563. YONGE, C. M. 1924. Studies on the comparative physiology of digestion. II, The mechanism of feeding, digestion and assimilation in *Nephrops norvegicus*. *J. exp. Biol.* **1** : 343–389.

564.   (Same as 563.)

565.   1932. On the nature and permeability of chitin. I, The chitin lining the foregut of decapod crustacea and the function of the tegumental glands. *Proc. R. Soc. B.* **111** : 298–329.

566.   1936. On the nature and permeability of chitin. II, Permeability of the uncalcified chitin lining of the foregut of *Homarus*. *Proc. R. Soc., B.* **120** : 15–41.

567. ZUCKERHANDYL, E. 1957. La teneur en oxygène de hémolymphe artérielle de *Maia squinada* aux divers stades d'intermue. *C. R. Soc. Biol.* **151** : 524–528.

Acknowledgements

We are grateful for permission to reproduce illustrations from various sources. Details of these sources are given in the list below.

In this book Fig. No.	Ref. No. of source	Source (book or periodical)	Author(s)	Publisher	In the source Vol. & Page	Fig. No.
		BOOKS				
1*A*	—	*Manual of Elementary Zoology*	Borradaile	O.U.P.		203, 210
1*B*	—	*Dana Reports* Vol. 5 (Report No. 27, 1945)	Hermann Einarsson	Carlsberg Foundation, Copenhagen		18
1*C*, D	—	*The Life of Crustacea*	Calman	Methuen		
1*E*	—	*Outlines of Zoology*	Thomson (revised Ritchie)	O.U.P.		*Praunus*
2*A-E*	—	*An Account of the Crustacea of Norway*	G.O. Sars	Bergen Univ. Publikasjons-utvalget		Plates 43, 177
3*A*, C, D, E	—	*Animal Life in Freshwater*	Mellanby	Methuen		35, 26, 64
3*B*	—	*The Seashore (New Naturalist Series)*	Yonge	Collins		31
		PERIODICALS				
4	440	*J. gen. Physiol*	Reuben et al.	Rockefeller Univ. Press	47: 1141-1194	2
5	320	*Comp. Biochem. Physiol.*	Lance	Pergamon	14: 155-165	
6, 7	340	*Biol. Rev.* (1962)	Lockwood	Camb. Phil. Soc.		3, 5
8, 9, 10, 11		*Animal Body Fluids and their Regulation*	Lockwood	Heinemann		4, 7
12	315	*Zool. Beiträge*	Kümmel	Duncker & Humblot	10: 227-252	3 Figs.
13	548	*Biol. Bull. Woods Hole*	Werntz	Marine Biol. Lab.	124: 225-239	2

In this book		Source (book or periodical)	Author(s)	Publisher	In the source	
Fig. No.	Ref. No. of source	PERIODICALS			Vol. & Page	Fig. No.
14	339	*J. exp. Biol.*	Lockwood	Co. of Biol. Ltd.	**38:** 647-658	1
15 *A*, B, C	137	*Proc. R. Soc. B.*	Dennell	The Royal Society	**134:** 485-503	2a
15 D	565	*Proc. R. Soc. B.*	Yonge	The Royal Society	**111:** 298-329	8, 9
16 A, B, D	94	*Proc. R. Soc. B.*	Carlise and Knowles	The Royal Society	**141:** 261	7
16 E	293	*Gen. compar. Endocrinol*	King	Academic Press, New York	**4:** 533-534	1
16 F	369	*Gen. compar. Endocrinol*	Maynard	Academic Press, New York	**1:** 237-263	12
17	43	*Biol. Bull., Woods Hole*	Bliss and Boyer	Marine Biol. Lab. Woods Hole	**103:** 20	2
18	55	*Biol. Bull., Woods Hole*	Brown and Jones	Marine Biol. Lab. Woods Hole	**96:** 228-232	1
19	297	*Biol. Bull., Woods Hole*	Kleinholtz	Marine Biol. Lab. Woods Hole	**70:** 165	1
20	298	*Biol. Bull., Woods Hole*	Kleinholtz	Marine Biol. Lab. Woods Hole	**72:** 30	1
21 A, B	—	*Liverpool Mar. Biol. Comm. Memoirs*	Hewitt	Liverpool Univ. Press, 123 Grove St., Liverpool, 7	**14**	1, 3
21 C	—	*Liverpool Mar. Biol. Comm. Memoirs*	Jackson	Liverpool Univ. Press, Liverpool, 7	**21**	Plate 4
22	14	*J. exp. Biol.*	Arudspragasam and Naylor	Company of Biologists, Cambridge	**41:** 299-308	7
23 A, B	437	*J. cell. comp. Physiol.*	Redmond	Wistar Inst. of Anat. & Biol., 36th Street at Spruce, Philadelphia, Pa.	**46:** 209-247	9, 4
23 C	435	*Biol. Rev.*	Redfield	Camb. Phil. Soc. London Cambridge Univ. Press	**9:** 196	7

In this book Fig. No.	Ref. No. of source	Source (book or periodical) PERIODICALS	Author(s)	Publisher	In the source Vol. & Page	Fig. No.
24	383	*J. exp. Biol.*	Needham	The Company of Biologists, 83 Downing St., Cambridge	**10**: 83	1
25	124	*Comp. Biochem. Physiol.*	Dean and Vernberg	Pergamon Press Inc.	**14**: 29	1
26	29, 348, 374, 532.	Schematic drawing based on the sources cited in Ref. list viz. *Comp. Biochem. Physiol*, and		Pergamon		
	411	the book: *The Liver* (ed. Rouiller)		Academic Press, New York.		
27	411	*The Liver* (ed. Rouiller)		Academic Press New York		
28	348	*Comp. Biochem. Physiol.*	McWhinnie & Corkill	Pergamon	**12**: 81-93	2
29	503	*Physiological Zoology*	Teal	University of Chicago Press, 5750 Ellis Avenue, Chicago, 37	**32**: 8	3
30	453	*Physiol. Zoology*	Roberts	,,	**30**: 243- 255	2
31	549	*Physiol. Zoology*	Weymouth et al.	,,	**17**: 50-71	1
32	520	*Biol. Bull. Woods Hole*	Vernberg	Marine Biol. Lab. Woods Hole, Mass.	**117**: 582- 593	5
33	349	*Comp. Biochem. Physiol.*	McWhinnie and Kirchenberg	Pergamon	**6**: 121	1
34 A, B, C	58	*J. cell. comp. Physiol.*	Brown, Bennett & Webb	Wistar Institute of Anat. & Biol., 36th Street at Spruce, Philadelphia, Pa.	**44**: 501	6A, B, C
35 A, B	344	*J. exp. Biol.*	Lofts	The Company of Biologists	**33**: 734	(both parts)
36	507	*J. exp. Biol.*	Thomas	The Comp. of Biologists	**31**: 242	11
37	551	*Physiol. Zoology*	Wiens & Armitage	University of Chicago Press	**34**: 44	2
38 A	261	*J. Physiol.*	Hoyle & Wiersma	Cambridge Univ. Press	**143**: 404	1

In this book Fig. No.	Ref. No. of source	Source (book or periodical)	Author(s)	Publisher	In the source Vol. & Page	Fig. No.
		PERIODICALS				
38 B	557	*Physiologia comp. et Oecol.*	Wiersma & Ripley	Dr. W. Junk, 13 Van Stolk- weg, The Hague	**2**: 392	1
40 A-G	261	*J. Physiol.*	Hoyle and Wiersma	C.U.P.	**143**: 403- 425	1, 8, 3, 13, 8
41	178	*J. Physiol.*	Fatt and Katz	C.U.P.	**120**: 175	2
42	156	*J. Physiol.*	Dudel and Kuffler	C.U.P.	**155**: 545	2
43	390	*J. Physiol.*	Orkand	C.U.P.	**161**: 148 (adapted)	2
44	272	*J. Physiol.*	Huxley and Taylor	C.U.P.	**144**: 426- 441	Drawn from text descrip- tion
45	268	*J. exp. Biol.*	Hughes and Wiersma	Company of Biologists	**37**: 293	2
46	33	*Archiv. mikroscop. Anat. EntwMech. I*	Bethe	Springer Verlag, Heidelberg/ Berlin	**50**: 330	369, 370
48	556	*J. comp. Neurol.*	Wiersma and Hughes	Wistar Insti- tute of Anat. & Biol., 36th St. at Spruce, Philadelphia, Pa.	**116**: 209- 228	11
49	268	*J. exp. Biol.*	Hughes and Wiersma	The Company of Biol.	**37**: 292	1
50	267	*J. exp. Biol.*	,,	,,	**37**: 295	3
51 A	282	*J. cell. comp. Neurol.*	Johnson	Wistar Insti- tute of Anat. & Biol., 36th St. at Spruce, Philadelphia, Pa.	**36**: 323- 373	(adapted from John- son)
51 B	213	*J. Physiol.*	Furshpan and Potter	C.U.P.	**145**: 289 325	1
52	556	*J. compar. Neurol.*	Wiersma and Hughes	Wistar In- stitute of Anat. & Biol. 36th Street at Spruce, Phil- adelphia, Pa.	**116**: 209- 228	5, 6, 3, 10

In this book		Source (book or periodical)	Author(s)	Publisher	In the source	
Fig. No.	Ref. No. of source	PERIODICALS			Vol. & Page	Fig. No.
53 A, B, C, D	2	*Q. Jl. microsc. Sci.*	Alexandrowicz	Comp. of Biologists	**92**: 172, 169, 190	3D, 2D 11
54 B, C	167	*J. gen. Physiol.*	Edwards & Hagiwara	Rockefeller Univ. Press	**43**: 315- 321	3 (adap- ted)
54 D	236	*J. gen. Physiol.*	Hagiwara & Naka	Rockefeller Univ. Press	**48**: 141- 162	Drawn from Text des- cription
	312	*J. gen. Physiol.*	Kuffler and Eyzaguirre	Rockefeller Univ. Press	**39**: 155- 184	Drawn from Text des- cription
55	3	*Q. Jl. microsc. Sci.*	Alexandrowicz	Company of Biologists	**93**: 332	9
56 A, C	550a	*Phil. Trans. R. Soc. B.*	Whitear	The Royal Society	**245**: 291- 324	11
56 B	78	*J. exp. Biol.*	Bush	Company of Biologists	**39**: 90	1
57 A	375	*J. exp. Biol.*	Mellon	Company of Biologists	**40**: 139	1
57 B	326	*Comp. Biochem. Physiol.*	Laverack	Pergamon	**5**: 320	2
57 C	327	*Comp. Biochem. Physiol.*	Laverack	Pergamon	**6**: 138	1
57 D	329	*Comp. Biochem. Physiol.*	Laverack	Pergamon	**10**: 270	9
58 A	105	*J. Physiol.*	Cohen	Journal of Physiol., Physiol. Lab., Downing St., Cambridge	**130**: 13	2
58 B	105	*J. Physiol.*	Cohen	,,	**130**: 18	6
59	105	*J. Physiol.*	Cohen	,,	**130**: 18	7
60 A, B	172	*Comp. Biochem. Physiol.*	Enright	Pergamon	**7**: 136	5, 1
61 A, B	314	*Symp. Soc. exp. Biol.*	Kuiper	The Comp. of Biologists	**16**: 59 64 65	1, 7, 8 (adap- ted)

In this book		Source (book or periodical)	Author(s)	Publisher	In the source	
Fig. No.	Ref. No. of source	PERIODICALS			Vol. & Page	Fig. No.
62 A, B	314	*Symp. Soc. exp. Biol.*	Kuiper	The Comp. of Biologists	**16:** 66	9
63 A	75	*Symp. Soc. exp. Biol.*	Burkhardt	The Company of Biol.	**16:** 89	3
63 B, C	83	*J. cell. comp. Physiol.*	Camougis	Wistar	**63:** 349, 351	7, 8
64	529	*Nature*	Wald and Hubbard	Macmillan, London	**180:** 279	1 (simplified)
65	330	*Comp. Biochem. Physiol.*	Laverack	Pergamon	**13:** 305	4
66 A	—	*Manual of Zoology*	Borradaile	O.U.P.	203	210
66 B	564	*J. exp. Biol.*	Yonge	Company of Biologists	**1:** 355	5
67	117	*J. exp. Biol.*	Croghan	Company of Biologists	**35:** 243– 249	1
68	294	*Veröff. Inst. Meeresforsch. Bremerh.* reproduced in *Z. wiss Zool.*	Kinne	Akademische Verlagsges., Geest and Portig, Leipzig	**6:** 177– 202. **157:** 427– 491	3

Generic Index

Numbers in *italic* refer to figures.

General Index

Numbers in *italic* refer to figures. The letters ff after an entry indicate detailed discussion of the topic on following pages.